Control of Water Quality Deterioration Caused by Corrosion of Cement-Mortar Pipe Linings

Prepared by:
Bruce D. Douglas and
Douglas T. Merrill
Brown and Caldwell Consultants
Pleasant Hill, CA 94523

Prepared for:
AWWA Research Foundation
6666 West Quincy Avenue
Denver, CO 80235

Published by the
AWWA Research Foundation and
American Water Works Association

Disclaimer

ISBN 0-89867-600-2

Printed on recycled paper.

Control of Water Quality Deterioration Caused by Corrosion of Cement-Mortar Pipe Linings

Contents

List of Tables vii

List of Figures viii

Foreword ix

Acknowledgments xi

Executive Summary xiii

Chapter 1 Introduction 1

Purpose of the Study, 1
Organization of the Report, 2

Chapter 2 Literature Review 3

Literature Review Methodology, 3
Cement and Mortar Characteristics, 3
  Cement Chemistry, 4
  Mortar Aggregates, 5
  Additives, 5
History of Cement-Mortar Lining, 6
Lining Methods, 6
  In Situ Lining, 6
  Factory Lining, 7
  Field Lining, 8
Cement-Mortar Lining Corrosion, 8
  US Studies, 9
  German Studies, 10
  British Studies, 11

Chapter 3 Utility Survey 13

Survey Methodology, 13
Results, 15
  Significant Water Quality Problems, 16
  Minor Water Quality Problems, 18
  Structural Problems, 18
Summary and Conclusions, 19

Chapter 4 Field Test Matrix and Procedures 21

Test Apparatus, 21
Test Matrix, 24

Operating Procedures, 28
  Sampling, 28
  Analysis, 29
Post-Test Lining Inspections, 31
  Appearance, Loose Sand, and Hardness, 31
  Scale Composition, Calcium Loss, and pH Effects, 31

Chapter 5 Field Test Results 33

Water Quality Effects, 33
  South Central Connecticut RWA Water Quality Results, 35
    Inlet Water, 35
    Flow Regime, 35
    Drag Trowel Versus Centrifugal Lining, 37
    Seal Coat, 41
    Water-Conditioning Chemicals, 41
    Mortar Composition, 42
    Other Inorganic Water Quality Parameters, 43
  Marin MWD Water Quality Results, 45
    Inlet Water, 45
    Flow Regime, 45
    Drag Trowel Versus Centrifugal Lining, 51
    Seal Coat, 51
    Water-Conditioning Chemicals, 52
    Mortar Composition, 52
    Other Inorganic Water Quality Parameters, 53
    Organic Water Quality Parameters, 53
Lining Effects, 58
  Physical Inspection, 58
  Calcium Loss, 60
  pH Effect, 62
Examination of Aged Pipe Sections, 62
  Physical Inspection, 63
  Scale Analyses, 63
  Calcium Loss, 63
  pH Effect, 65
Summary, 65

Chapter 6 Conclusions and Recommendations 69

References 71

List of Abbreviations 73

Appendix A Field Test Results 75

Tables and Figures

Tables

2.1	Basic composition of Type I portland cement	4
3.1	Surveyed utilities	14
3.2	Survey results—Utility satisfaction and problems with cement-mortar lining	15
3.3	Finished water quality data for utilities with significant pH increases	16
4.1	Test matrix	25
4.2	Characteristics of hardened cement mortars used for in situ lining	27
4.3	Sampling and analysis during first two weeks	29
4.4	Sampling and analysis during final 11 weeks	30
4.5	Field test sections analyzed for scale composition, calcium loss, and pH effects	32
5.1	Summary of South Central Connecticut RWA field test results	36
5.2	South Central Connecticut RWA inlet versus continuous-flow water quality	37
5.3	South Central Connecticut RWA inlet water quality for other inorganic parameters	43
5.4	Summary of Marin MWD field test results	46
5.5	Marin MWD inlet versus continuous-flow water quality	47
5.6	Marin MWD inlet water quality for other inorganic parameters	53
5.7	Marin MWD organic compound concentrations (µg/L)	56
5.8	Composition of scales or other portions of linings for test sections (percent)	59
5.9	Scale analyses for sections A, B, and C (percent)	64

Figures

4.1 Conceptual diagram of pipe section array 22

4.2 South Central Connecticut RWA field test rig 22

4.3 Marin MWD field test rig 23

5.1 South Central Connecticut RWA pH results 38

5.2 South Central Connecticut RWA alkalinity results 39

5.3 South Central Connecticut RWA calcium results 40

5.4 South Central Connecticut RWA silica results 44

5.5 Marin MWD pH results 48

5.6 Marin MWD alkalinity results 49

5.7 Marin MWD calcium results 50

5.8 Marin MWD silica results 54

5.9 Calcium content (as CaO) versus average mortar depth for reference mortar for three different flow regimes—South Central Connecticut RWA 61

5.10 Calcium content (as CaO) versus average mortar depth for reference mortar for three different flow regimes—Marin MWD 61

5.11 pH of slurries made by mixing cuttings from different depths with deionized water—South Central Connecticut RWA 62

5.12 Calcium content (as CaO) versus average mortar depth for various exposure times—South Central Connecticut RWA 64

5.13 pH of slurries made by mixing cuttings from different depths with deionized water—South Central Connecticut RWA 66

Foreword

The AWWA Research Foundation is a nonprofit corporation that is dedicated to the implementation of a research effort to help local utilities respond to regulatory requirements and traditional high-priority concerns of the industry. The research agenda is developed through a process of grass-roots consultation with members, utility subscribers, and working professionals. Under the umbrella of a Five-Year Plan, the Research Advisory Council prioritizes the suggested projects based upon current and future needs, applicability, and past work; the recommendations are forwarded to the Board of Trustees for final selection.

This publication is a result of one of those sponsored studies, and it is hoped that its findings will be applied in communities throughout the world. The following report serves not only as a means of communicating the results of the water industry's centralized research program but also as a tool to enlist the further support of the nonmember utilities and individuals.

Projects are managed closely from their inception to the final report by the Foundation's staff and large cadre of volunteers who willingly contribute their time and expertise. The Foundation serves a planning and management function and awards contracts to other institutions such as water utilities, universities, and engineering firms. The funding for this research effort comes primarily from the Subscription Program, through which water utilities subscribe to the research program and make an annual payment proportionate to the volume of water they deliver. The program offers a cost-effective and fair method for funding research in the public interest.

A broad spectrum of water supply issues is addressed by the Foundation's research agenda: resources, treatment and operations, distribution and storage, water quality and analysis, toxicology, economics, and management. The ultimate purpose of the coordinated effort is to assist local water suppliers to provide the highest possible quality of water economically and reliably. The true benefits are realized when the results are implemented at the utility level. The Foundation's Trustees are pleased to offer this publication as a contribution toward that end.

Cement-mortar lining has been an effective means of protecting iron and steel water pipelines from the corrosive action of water. Some utilities have suspected cement-mortar lining as a cause for deterioration of water quality, particularly in pipelines that are lined in the field (in situ). This study evaluates the impact of commonly used cement-mortar mixes on water quality at two different drinking water utilities. This study discusses measures a utility can implement to minimize water quality deterioration caused by cement-mortar pipe linings.

Richard P. McHugh
Chair, Board of Trustees
AWWA Research Foundation

James F. Manwaring, P.E.
Executive Director
AWWA Research Foundation

Acknowledgments

Many people contributed to the success of this project. The authors wish to thank Joel Catlin, AWWARF project officer, and the members of the project advisory committee—Chris Hebberd, South Central Connecticut Regional Water Authority; Dave Heumann, Los Angeles Department of Water and Power; and Stephen McGrew, Palm Beach County Water Utilities Department. We thank Richard Thomasson of the Washington Suburban Sanitary Commission for securing the Commission's financial assistance for the project.

In addition, we appreciate the efforts of the people who assisted with the two utility field tests. From South Central Connecticut Regional Water Authority, we extend our gratitude to Tom Kalinosky, Elizabeth Paluba, and Noel Grant. We also thank Allen Brown and Ken Cuneo of Marin Municipal Water District.

We also appreciate the efforts of John Hohider of Heitkamp, Inc., who supervised construction of many of the pipe test sections, Bob Haffner of Calcoast Analytical Laboratory, who chemically analyzed the pipe linings after completion of the field tests, and Richard Bonds of the Ductile Iron Pipe Research Association, who provided helpful comments to the draft report.

Executive Summary

Introduction

Ferrous metal water distribution pipe can corrode by electrochemical action, which leads to scale, leaks, and tuberculation. Lining the pipe interior with cement mortar protects the metal from corrosion by providing a coating of electrically inert material between metal and water. However, the question arises, does the lining itself tend to corrode? "Corrosion," in this sense, means the leaching of lime from the cement matrix, causing calcium loss and possibly leading to structural deterioration of the lining. In addition, the leached lime can impair water quality by increasing the pH, calcium, and alkalinity, which may result in customer complaints.

Objectives

The purpose of this project was to study the causes and effects of cement-mortar lining corrosion. Principal objectives were to:

- Investigate the extent of the corrosion problem in the United States.
- Identify factors that lead to corrosion.
- Examine the effects of various water quality parameters on commonly used cement mortars.
- Recommend measures a utility can implement to avoid corrosion problems.

The report focuses on in situ (in place) pipe lining, because few problems have been noticed with standard seal-coated and factory-lined pipe. However, factory-lined and seal-coated pipe are also discussed.

Methods

The study includes a literature review, utility survey, and field tests. The literature indicated that cement-mortar lining corrosion is most likely to occur in newly lined pipe carrying low-alkalinity water. Based on this information, the survey concentrated on utilities that obtain water from rainfall or snowmelt runoff, which is generally less alkaline than groundwater. About 10 percent of the 33 surveyed utilities experienced cement-mortar lining corrosion that caused serious water quality problems—most notably, pH increases to 9 or above. Corrosion effects (for example, degraded water quality) were most noticeable in dead ends or where the flow is low or intermittent. These problems required many months of flushing to correct.

Two of the surveyed utilities subsequently participated in the project by conducting three-month pilot-scale field tests to determine the condition(s) leading to cement-mortar lining corrosion. Each field test operated with 10 lined pipe sections that differed from each other in one parameter only. This was done in order to isolate the effect of each parameter. The parameters included water flow regime, lining method, seal coating, water quality, and mortar composition.

Results

The field tests showed that water can leach cement compounds from linings, causing the water's pH, alkalinity, and calcium values to increase. Water that remained in pipe sections for a long time (longer than 1 week) exhibited significant water quality effects—for example, pH increases from an initial value of 7 up to 12. If the water flowed continuously through a pipe section, the constant flushing reduced water quality impacts by dilution but leached more cement compounds from the lining, as indicated by the depth of the lining's calcium loss. None of the linings appeared to be structurally weakened during the tests.

Aggressive water promoted leaching. At the field test sites, waters with lower Langlier Indexes had larger pH, alkalinity, and calcium increases.

Asphaltic seal coats significantly reduced cement-mortar corrosion in both field tests, based on smaller pH, alkalinity, and calcium increases. To investigate if the seal coat leached organic compounds, one field test utility analyzed the water from the seal-coated section for priority organic pollutants during the last week of the study. The analyses identified only one compound, tetrachloroethylene (PCE), in unexpectedly high concentrations. This unexpected result may have been an isolated incident since PCE leaching has apparently not been discovered in tests conducted by others. Further research is needed to investigate whether the seal coat is a potential source of PCE or other priority organic materials.

The effects of mortar composition do not necessarily translate from one utility to another. At one field test site, corrosion increased when the mortar contained Type II cement, a water-reducing admixture, pozzolanic material, or a higher proportion of sand. At the other site, none of these parameters had a significant impact on results.

In summary, utilities with aggressive water should expect cement-mortar lining corrosion to increase the water's pH, calcium, and alkalinity values in newly lined pipe, particularly in low-flow areas. These effects will generally be less with seal-coated pipes. Systems should be designed with the most corrosion-resistant mortars. Utilities should regularly flush dead ends and low-flow areas to prevent buildup of components causing adverse water quality effects.

Chapter 1

Introduction

Ferrous metal water distribution pipe can corrode by electrochemical action, which leads to scale, leaks, and tuberculation. Lining the pipe interior with cement mortar protects the metal from corrosion by providing a coating of electrically inert material between metal and water. The procedure for rehabilitating corroded pipe by in situ lining was developed in the 1930s and has evolved into a widely practiced corrosion control measure.

While cement-mortar lining has been successful in protecting against electrochemical corrosive effects, recent research indicates that the lining itself may not be immune to corrosion. "Corrosion," in this sense, means the leaching of lime from the mortar, causing calcium loss and possibly leading to structural deterioration of the lining. In addition, leached lime may impair water quality by increasing the pH, alkalinity, and calcium concentrations.

Corrosion appears to be most likely to occur in newly lined pipes carrying low-alkalinity water (Warren and Elvidge 1988). Some cement-mortar linings, after many years of service, have been found to be sandy or roughened in texture and show evidence of low cohesive strength, possibly caused by exposure to aggressive water (Miller 1965). A few utilities have reported problems possibly caused by high pH (taste or skin irritation) that have lasted for more than one year after lining pipes with cement mortar in low-flow areas, such as dead ends. In contrast, other utilities have used cement-mortar-lined pipes for many years without lining deterioration or customer complaints. The reasons for problems occurring at one utility and not at another are not well known.

Purpose of the Study

The purpose of this project was to study the causes and effects of cement-mortar lining corrosion. Principal objectives were to (1) investigate the extent of the corrosion problem in the United States, (2) identify factors leading to corrosion, (3) examine the effects of various water quality and cement matrix parameters on cement-mortar corrosion, and (4) recommend measures a utility can implement to minimize corrosion problems. The report focuses on in situ (in place) pipe lining, because few problems have been noticed with standard seal-coated and factory-lined pipe. However, factory-lined and seal-coated pipe are also discussed.

Organization of the Report

This report is organized as follows:

- Chapter 1 introduces the problem of cement-mortar lining corrosion.
- Chapter 2 summarizes the literature regarding cement-mortar lining, including lining history, techniques, and corrosion studies.
- Chapter 3 presents a survey of utility experiences with cement-mortar lining, gauging the extent of the corrosion problem and the types of problems encountered.
- Chapters 4 and 5 describe field tests conducted at two utilities to determine how water flow regime, lining method, seal coating, water quality, and mortar composition affect lining corrosion. Chapter 4 explains the test matrix, fabrication of lined pipe sections, and test procedures. Chapter 5 evaluates the results of the field tests.
- Chapter 6 summarizes the conclusions of the project and recommends measures for minimizing cement-mortar lining corrosion problems.

Chapter 2

Literature Review

For more than 60 years, cement-mortar lining has been an effective means of protecting iron and steel water pipelines from the corrosive action of water. However, recent studies and utility experience have raised the question of whether cement-mortar lining itself tends to corrode. This chapter reviews the literature on cement-mortar lining corrosion.

Literature Review Methodology

Computerized bibliographic databases were consulted via the extensive Dialog Network* to access information written in English from the United States and other countries. First, the searches scanned the databases using broad-based key words, such as "cement-mortar lining" and "pipeline rehabilitation." These searches produced nearly 100 titles of articles, books, and conference proceedings. The second step was to print out abstracts for the promising titles. Third, after examining the abstracts, the publications that addressed cement-mortar lining corrosion were obtained and studied. This three-step process minimized the duplication of information and controlled costs.

Much of the pertinent research carried out in Germany, Holland, France, and Japan is not available in English. Only publications written in English were reviewed.

The remainder of this chapter presents the results of the literature review. The characteristics of cement and cement mortar are described, the history of cement-mortar lining is reviewed, and lining methods are described. Finally, previous studies of cement-mortar lining corrosion in the United States, Germany, and Great Britain are discussed.

Cement and Mortar Characteristics

Cement mortar is formed by mixing cement, mortar aggregates, water, and if desired, other additives. When used for pipeline rehabilitation, a typical mortar contains roughly equal weights of cement and aggregate (sand), with sufficient water added to form a workable mixture. Additive proportions can vary considerably, depending on the purpose of the additive; weight proportions up to 10 percent are

*Dialog Information Services, Inc., Palo Alto, Calif.

common. This section describes cement chemistry, mortar aggregates, and the use of additives. Much of the information was taken from *Composition and Properties of Concrete*, by Troxell et al. (1968).

Cement Chemistry

Cement is a finely powdered substance that forms a strong, durable material when mixed with water and allowed to harden. While its exact composition is not fully understood or identified, it is customary to define the principal cement compounds as oxides of calcium, silica, aluminum, and iron. Table 2.1 presents the typical composition of portland Type I cement, the most common cement for ordinary use (JM Montgomery 1985). This breakdown shows that tricalcium silicate and dicalcium silicate account for the bulk of cement materials. Other important compounds may include tricalcium aluminate, tetracalcium alumino ferrite, calcium sulfate, magnesium oxide, and free lime.

When cement is mixed with water to form a paste, a complex series of reactions begins, in which the compounds hydrate, go into solution, and form new hydrated precipitates (Troxell et al. 1968). Particularly important are the reactions of the calcium silicates:

- For tricalcium silicate:

$$2(3CaO \cdot SiO_2) + 6H_2O \rightarrow 3CaO \cdot 2SiO_2 \cdot 3H_2O + 3(CaO \cdot H_2O)$$

- For dicalcium silicate:

$$2(2CaO \cdot SiO_2) + 4H_2O \rightarrow 3CaO \cdot 2SiO_2 \cdot 3H_2O + (CaO \cdot H_2O)$$

Each of these reactions forms a hydrated calcium silicate, the basic gel structure that characterizes hardened cement paste.

The above calcium silicate reactions release appreciable quantities of lime ($CaO \cdot H_2O$). Additional lime goes into solution from the hydration of other cement

Table 2.1 Basic composition of Type I portland cement

Compound	Formula	Cement composition (%)
Tricalcium silicate	$3CaO \cdot SiO_2$	42–67
Dicalcium silicate	$2CaO \cdot SiO_2$	8–31
Tricalcium aluminate	$3CaO \cdot Al_2O_3$	5–14
Tetracalcium alumino ferrite	$4CaO \cdot Al_2O_3 \cdot Fe_2O_3$	6–12
Calcium sulfate	$CaSO_4$	2.6–3.4
Free lime	CaO	0–1.5
Magnesium oxide	MgO	0.7–3.8
Volatiles	—	0.6–2.3

compounds and from free lime that simply dissolves. Dissolved lime releases calcium and hydroxyl ions, as shown below:

$$CaO \cdot H_2O \rightarrow Ca^{2+} + 2OH^-$$

This reaction increases the water's calcium, pH, and alkalinity. Under the resulting alkaline conditions, the calcium ions may combine with carbonate ions in the water to form calcium carbonate at the lining surface (Holtschulte and Schock 1985). The calcium carbonate precipitate may coat the lining and hinder further lime dissolution. However, if the cement contacts aggressive water (low pH or alkalinity), the $CaCO_3$ coating cannot form. Then lime may leach excessively from the mortar, leading to cement-mortar corrosion.

Mortar Aggregates

The functions of the aggregate are (1) to provide a relatively inexpensive filler for the mortar; (2) to provide a mass of particles suitable for resisting the action of applied loads, abrasion, and the percolation of moisture; and (3) to reduce the volume changes resulting from the setting and hardening process and from moisture changes in the cement–water paste (Troxell et al. 1968).

Sand is the aggregate used in cement-mortar linings. A clean, well-graded sand is important for preparing a workable mortar paste that will deliver proper strength.

Additives

Various compounds can be added to the cement mortar to produce certain desirable qualities. Additives can reduce the water required, promote pozzolanic reactions, entrain air, improve bonding characteristics, aid curing, and accelerate setting. The two additives considered in this report are water-reducing admixtures and pozzolans.

Water-Reducing Admixtures

A given cement-mortar paste has a certain workability, measured by its slump.* A water-reducing admixture allows the paste to achieve the same workability with less water. A lower water content produces a denser, less porous cement mortar that, upon setting, has a higher compressive strength (Troxell et al. 1968). Such a hardened mortar absorbs less water and, therefore, may be less prone to corrosion from lime leaching.

Pozzolans

A pozzolan is a silica-based material that reacts with lime at ordinary temperatures in the presence of water to form calcium silicate. Therefore, a pozzolan

*In the slump test, mortar is packed into a mold shaped like a truncated cone, 12 in. (inches) in height, with a diameter decreasing from 8 in. at the base to 4 in. at the top. When the mold is lifted off, the mortar pile subsides. The amount of subsidence, in inches, is the slump. ASTM C143 (1990a) describes the slump test.

can tie up lime produced during cement hydration reactions, forming strength-producing compounds similar to desirable cement end products. This reaction could prevent lime from leaching and the pH from increasing unduly. Portland-pozzolan cements gain strength relatively slowly, but may ultimately attain greater strength than a nonpozzolan cement (Troxell et al. 1968).

The primary source of artificial pozzolans is fly ash, the glassy residue that results from the combustion of pulverized coal at electric power plants. Typical fly ash is made up of silica and alumina oxides, with some iron and other oxides. The major reaction of fly ash and lime is (Smith and Raba 1980):

$$SiO_2 + Al_2O_3 + Ca(OH)_2 + H_2O \rightarrow xCaO \cdot yAl_2O_3 \cdot SiO_2 \cdot wH_2O$$

History of Cement-Mortar Lining

As long ago as 1836, the French Academy of Sciences recommended that cement lining be used to protect the interior of cast-iron water pipes from corrosion. In 1843, a method for lining pipe above ground was patented by Jonathan Ball (Speller 1951). By the middle of the 1800s, many small towns in the New England area, where the water was characterized as very soft and slightly acidic, were using cement-lined pipes for water services (Miller 1965).

During the 1870s, a type of cement-lined and cement-wrapped pipe was manufactured and used in the New England area and the mid-Atlantic states. This pipe consisted of a very thin wrought-iron shell that was both wrapped and lined with about ½ in. (inch) (13 mm [millimeters]) of cement. The useful life of this pipe was about 25 to 30 years. In 1922, the first cement-lined, cast-iron pipe was installed in Charleston, SC (Miller 1965).

Practical methods for in situ lining were developed in the United States beginning in the 1930s. The first field trial of in situ lining took place in 1933 to rehabilitate a 72-in. (1,830-mm) pipe section in Jersey City, NJ. Cement-mortar lining methods, both in situ and at the factory, were incorporated into AWWA standards in 1939. Improved methods and techniques currently allow in situ lining of pipe as small as 4 in. (100 mm) in diameter (AWWA 1989a). Cement-mortar lining is now widely used as a means for rehabilitating and protecting water pipelines.

Lining Methods

Cast-iron, ductile iron, and steel pipe can be lined in situ (as a means of rehabilitation), at the factory, or in the field prior to installation. Lining methods for each approach are described below.

In Situ Lining

In situ lining is used to rehabilitate an existing underground pipe that suffers from metallic corrosion. It has largely been used to rehabilitate old, unlined cast-iron pipe. It has not, to the author's knowledge, been used to rehabilitate factory-lined pipe. *AWWA Standard for Cement-Mortar Lining of Water Pipelines—4 in. (100 mm) and Larger—In Place* (ANSI/AWWA Standard C602) describes an industry standard for in situ lining. Three in situ methods are described below: drag trowel, projection, and poured-in-place.

Before beginning the lining procedure, workers thoroughly clean the targeted pipe section and mix the cement mortar in an aboveground hopper. The most common mixture for in situ lining is one part fine silica sand to one part portland cement.

Drag Trowel Method

In the drag trowel method of application, a slug of mortar is pressed into one end of the pipe. A tapered brass mandrel is then pulled through the pipe by a wire. The mandrel is long and cone-shaped, with the last 6 to 8 in. (150 to 200 mm) about 1⁄16 in. (1.6 mm) larger in diameter to help keep the coating even throughout the pipe (Speller 1951).

Typical lining thicknesses are from 0.13 to 0.50 in. (3 to 13 mm); the actual thickness of the lining depends on the pipe diameter, condition, and engineering judgment (Dahl 1982). Larger pipe diameters require thicker linings.

Projection Method

In the projection method, the mortar is injected into an application hose by a positive displacement pump. A spinning, air-driven spray head applies the mortar, backing away as the mortar is applied down the length of the pipe. Mechanical trowels smooth the finished surface. Curing typically takes 16 to 24 hours (Dahl 1982; Warren and Elvidge 1988; Radziul and Jackson 1965).

Poured-in-Place

For large water pipes (for example, greater than 6 ft (feet) (2 m [meters]) in diameter), forms may be built inside the pipe and the cement poured through temporary holes in the top of the pipe forms. This method has been used successfully without reinforcement (Speller 1951). The minimum curing time is 8 hours, but 16 to 24 hours is preferable (Radziul and Jackson 1965).

Factory Lining

Nearly all ductile iron pipe manufactured in the United States for drinking water service is cement-mortar lined at the factory. The most common process is the centrifugal method, governed by the *American National Standard for Cement-Mortar Lining for Ductile-Iron Pipe and Fittings for Water* (ANSI/AWWA Standard C104/A.21.4).

Centrifugal Process

Pipe can be lined prior to installation by the centrifugal process. The pipe is set on a bed of rollers, then a lance distributes cement along the length of the pipe while the pipe is slowly rotated at a rate of about 200 lfpm (linear feet per minute) (1 lmps [linear meter per second]). Once the necessary quantity of cement has been distributed evenly throughout the pipe, the rotation speed is increased to 600 lfpm (3 lmps) or more for about 30 seconds (Speller 1951; Babbitt and Doland 1949). The pipe is also vibrated during high-speed rolling to increase lining density (Bonds 1989). This high-speed spinning brings excess water and fine particles to the surface. A water wash is used to flush them from the surface, before the seal coat is applied (Bonds 1989). Factory-lined pipes usually have sand-to-cement ratios in the range of 1:2 to 1:3. Curing time for this method is 7 days. If hardening of the mortar is accelerated through heat treatment, then the curing time can be reduced to 2 days (Holtschulte and Schock 1985).

Pipe manufacturers normally apply a seal coat of asphaltic material to factory-lined pipe (unless specifically requested by the purchaser to leave it off). If applied immediately after lining, the seal coat can help control the rate of cement-mortar curing. Seal coating may also inhibit the corrosion of cement-mortar linings, because it creates a secondary layer of material between water and lining (Bonds 1989). The seal coat must not impair water color, taste, or odor. Furthermore, there are limits on the amount of extractable organic compounds that are allowed to leach from the seal coat. Extraction with distilled water for 24 hours must not yield chloroform-soluble compounds in excess of 0.028 mg/mm^2 (milligrams per square millimeter) or in excess of 50 mg/L (milligrams per liter) in the extracting water (AWWA 1990). Seal-coat manufacturers can apply to organizations such as the National Sanitation Foundation (NSF) for certification that a seal-coat formulation passes certain leaching criteria.

Projection Method

The projection method for factory-lined pipe is similar to the projection methods for in situ-lined pipe in that mortar is applied by a rapidly revolving head that has been inserted into a stationary pipe. However, in the factory projection method, the pipe is slowly rotated and vibrated once the mortar has been applied. The rotation/vibration helps smooth the mortar surface. The rotational role is not high enough to bring excess water and fine particles to the surface; thus, there is no need to wash the lining, as there is in the high-speed centrifugal process (Bonds 1989).

Field Lining

Steel pipe may be lined at a field facility prior to installation. These facilities use the centrifugal method described in *AWWA Standard for Cement-Mortar Protective Lining and Coating for Steel Water Pipe—4-in. and Larger—Shop Applied* (ANSI/AWWA C205).

Cement-Mortar Lining Corrosion

An increase in the water pH, calcium, and alkalinity typically follows initial contact with cement-mortar lining. The increase is caused by dissolving lime and may last as long as 1 to 4 years (Uhlig 1948; Speller 1951; Holtschulte and Schock 1985; Ashworth 1984). This is the natural result of the hydration and dissolution reactions that occur when cement contacts water. The assessment by a utility that this corrosion is a problem depends on the severity of the water quality changes, the duration of the changes, the occurrence of structural damage to the lining, and the extent of utility resources needed for corrective measures.

Uneven lining thickness or incomplete coverage of the surface can accelerate cement-mortar lining corrosion. The following conditions can cause these problems:

- Riveted pipe.
- Extremely flexible pipe.
- Improperly bedded pipe.
- Misaligned pipe.
- Pipe with internal defects, such as bulges caused by weld seams.

Water quality has a pronounced effect on the corrosion of cement-mortar linings. Some important water quality parameters are pH, alkalinity, carbon dioxide, calcium, silica, and possibly chloride, sulfate, zinc, magnesium, and manganese concentrations, as well as temperature (Holtschulte and Schock 1985).

West German Standard W342 (DVGW 1978) identifies the type of water for which cement-mortar lining is appropriate. Waters should have a total carbonic acid concentration of less than 0.25 mol/m^3 (gram–moles per cubic meter) (approximately 11 mg/L) as CO_2 and a calcium concentration of more than 0.02 mol/m^3 (approximately 1 mg/L) as Ca. Calcium leaching increases as the concentration of dissolved carbon dioxide increases. Furthermore, for pipes carrying waters classified as "calcium-dissolving," meaning a pH less than 7.8 and a Langlier Index (LI) less than -0.3, special precautions must be taken to ensure that the lining does not dry out at any time during the life of the pipe, including during repairs (Holtschulte and Schock 1985).

The pH of the cement mortar used in lining is important in protecting metal pipes carrying high-chloride water. Typically, cement-mortar linings, when first made up, have a pH of 12.6 or greater. Studies have shown that corrosion of metal pipe may result if the mortar pH is 12.5 and the water carried by the rehabilitated pipe has a chloride content of 700 mg/L. At a pH of 11.6, metal corrosion was noted at a chloride concentration of 71 mg/L. Mortar with a pH of 13.2 protected metal pipes to a chloride concentration of 8,900 mg/L (Benedict 1985).

Specific studies have been conducted to evaluate cement-mortar lining corrosion in the United States, Germany, and Great Britain. These studies are discussed below.

US Studies

As one measure of the corrosion of cement-mortar linings, the Hazen–Williams flow coefficient was measured for a number of pipes in 13 cities throughout the United States (Miller 1965). The Hazen–Williams coefficient is a commonly used measure of the roughness of the inner surface of a pipe; typical values range from about 75 to 140 (in dimensionless units), with a higher coefficient meaning a smoother pipe surface (Steel 1960). Pipes from 6 to 36 in. (150 to 915 mm) in diameter were tested. Testing was conducted on the pipes when new and again after different lengths of time in service (as much as 39 years after installation). It was found that a slight reduction in the Hazen–Williams flow coefficient occurred with time, typically 0.2 percent per year (Miller 1965).

In general, the linings successfully maintained high flow coefficients in the pipe. Linings without asphaltic seal coats, however, were subject to attack by aggressive water. These linings showed evidence of low cohesive strength after 30 years, but still gave adequate protection from tuberculation (Miller 1965).

German Studies

The AWWA Research Foundation and the DVGW-Forschungsstelle (DVGW-F), Karlsruhe, Germany, cooperatively prepared a comprehensive manual titled *Internal Corrosion of Water Distribution Systems*. One chapter (Holtschulte and Schock 1985) discusses the use of cement-mortar lining as a means of corrosion protection, the factors that influence corrosion of cement-mortar linings, and tests at several German utilities using cement-mortar lining. Key test results are summarized below.

Tests of cement-mortar-lined pipes were conducted in Dortmund, Hanover, Harz, and Fernwasserversorgung Bayerischer Wald (northeast of Bavaria). Both factory-lined and in situ-lined pipes were included in the study. It should be noted that the factory-lined and in situ-lined pipes were alike in that neither had an asphaltic seal coat applied to the lining.

Dortmund tests first took place in 1972. The water supply was characterized by a low pH (6.4 to 6.9), high sulfates (33.6 to 67.2 mg/L as SO_4), and high dissolved carbon dioxide (4.4 to 27.2 mg/L as CO_2). Factory- and in situ-lined pipes ranging from 3 to 40 in. (80 to 1,000 mm) were removed from service and tested. The tests included:

- Visual inspection.
- Determination of the mechanical stability of the lining.
- Total thickness of the lining versus the thickness of the hard portion of the lining.
- Calcium oxide, iron, and manganese concentrations in the coating on the lining.
- Percent of sand and mortar (by weight) versus specifications.
- Determination of the grain-size distribution of the sand.

The researchers found that for all cement-mortar linings, a layer formed on the water side of the lining. They speculated that calcium oxide had leached out of the lining but was replaced by hydroxides of iron and manganese. These hydroxides formed a gelatinous sediment that caused the cement-mortar layer to become very tight, thus reducing the rate of subsequent calcium oxide leaching. This phenomenon has been observed elsewhere (Carson 1927). Further, it was found that the factory-lined pipes all exhibited softening of the cement-mortar layer, whereas the in situ-lined pipes did not. In pipes with low velocity flows, no softening or leaching was observed. Increased concentrations of dissolved carbon dioxide accelerated the leaching rate.

Researchers found that, prior to contact with the water, factory linings had a higher lime concentration near the inner (water-side) surface than did the in situ linings. Furthermore, after water contact, a higher percentage of lime leached from the inner surface of the factory linings than from the surface of the in situ linings. They attributed these differences to the mechanisms of the two lining techniques. The centrifugal method, because of high-speed spinning, tends to segregate the sand from the mortar, forcing the heavier sand particles toward the outer portions of the lining (toward the metal pipe). The in situ method does not cause segregation. It places a uniform layer of mortar against the pipe. As a result, the water side is less rich in cement compared with centrifugally spun lining.

Additional testing at Dortmund, begun in 1980, involved pumping water of varying quality through lined pipes for 3 years. The study concluded that:

- A layer 0.04 to 0.16 in. (1 to 4 mm) thick on the water side of the lining was decalcified.
- The use of high-alumina cement and acrylic resin additives significantly increased the corrosion resistance of the mortar.
- Pipe that was lined at the factory by the centrifugal process was more resistant to corrosion than was pipe lined in situ (however, researchers questioned whether pipes lined above ground using in situ techniques truly represented normal underground in situ lining conditions).
- Water velocity had no influence on the rate of corrosion in tests where velocities ranged from 1 to 6.5 fps (feet per second) (0.3 to 2 mps [metres per second]).
- Less sand loss was experienced when lower water-to-cement ratios were used to prepare the mortar.
- Significant loss of sand only occurred when the lining was allowed to dry out and was brushed.

Addition of copolymers, such as acrylic resin or high alumina, improved the resistance of the lining to abrasion but showed little influence on the decalcification rate. The German Hygienic Institute at the University of Bonn found that the addition of acrylic resin to the cement-mortar lining was hygienically acceptable (Holtschulte and Schock 1985).

British Studies

The Water Research Centre, Buckinghamshire, Great Britain, reported problems with cement-mortar lining, including a significant reduction in flow-carrying capacity in small-diameter pipes and leaching of lime from the lining that can elevate the pH above the maximum advisable value of 9.5 (established by the Council of the European Communities 1980). The key water quality parameter associated with leaching was low alkalinity. It should be noted that these pipes were cement-mortar lined without an asphaltic seal coat.

Test sections of 4-in. (100-mm) diameter pipes located at dead-end portions of the distribution system were lined using ordinary portland cement and fine silica. Water alkalinities varied from 10 to 160 mg/L as $CaCO_3$. In the worst case, elevated pH values continued for over 2 years. Changing the mortar composition by substituting 65 percent blast furnace slag or fly ash for cement improved conditions somewhat, but the pH could still exceed 9.5 for several months. The Water Research Centre recommends an epoxy resin, not cement mortar, be used to line pipe carrying low-alkalinity water (Warren and Elvidge 1988).

Chapter 3

Utility Survey

US water utility personnel were surveyed by telephone to learn about their experiences with distribution pipes lined with cement mortar. The objectives of the survey were to:

- Determine the extent of cement-mortar lining corrosion problems in US water utility distribution systems.
- Identify specific types of corrosion problems.
- Select one or more utilities to participate in subsequent field testing.

The survey was conducted during the summer of 1989. This chapter describes the survey methodology, including the selection of contacted utilities and the survey questions, and then discusses survey results.

Survey Methodology

Because recent research indicates that most cement-mortar lining corrosion problems occur with low-alkalinity water, the survey concentrated on areas in the United States likely to have such water. Thus, the surveyed utilities were in New England; near the Appalachian, Sierra Nevada, Rocky, and Coast Range mountains; and in the Pacific Northwest. These areas obtain water largely from rainfall or snowmelt runoff, which is generally softer and less alkaline than groundwater.

Having selected the geographical regions to survey, specific utilities were identified using the AWWA 1988–1989 Membership Roster, professional committee membership listings, and client contacts. The survey contacts included utility officers, laboratory chemists, water plant staff, and field personnel. Table 3.1 lists the 33 surveyed utilities.

After explaining the purpose of the telephone call, the following questions were asked:

- Does the utility have cement-mortar-lined pipes in the distribution system?
- Has the utility experienced problems from corrosion of cement-mortar lining (for example, pH increase or lining deterioration)?
- Does the utility monitor the system for indications of cement-mortar lining corrosion (for example, analyzing distribution system pH or inspecting installed pipe)?

Table 3.1 Surveyed utilities

Utility Name	City	State
East Bay Municipal Utility District	Oakland	California
Marin Municipal Water District	Corte Madera	California
North Tahoe Public Utilities Department	Tahoe Vista	California
Palo Alto Water and Gas	Palo Alto	California
San Francisco Water Department	San Francisco	California
San Jose Water Department	San Jose	California
Tahoe City Public Utilities Department	Tahoe City	California
Truckee Donner Public Utilities Department	Truckee	California
Boulder Water Department	Boulder	Colorado
Colorado Springs Water Division	Colorado Springs	Colorado
Denver Water Board	Denver	Colorado
East Larimer County Water District	Fort Collins	Colorado
Grand Junction Water Department	Grand Junction	Colorado
Berlin Water Control Commission	Berlin	Connecticut
Danbury Public Utilities	Danbury	Connecticut
South Central Connecticut Regional Water Authority	New Haven	Connecticut
Washington Suburban Sanitary District	Laurel	Maryland
Boston Water and Sewer Commission	Boston	Massachusetts
Essex Water Department	Essex	Massachusetts
Holyoke Water Works	Holyoke	Massachusetts
Salem and Beverly Water Board	Beverly	Massachusetts
Carson City Water and Wastewater Utilities	Carson City	Nevada
Sierra Pacific Power Company	Sparks	Nevada
Concord Water Department	Concord	New Hampshire
Portland Bureau of Water Works	Portland	Oregon
Salem Public Works Department	Salem	Oregon
Greenville Water System	Greenville	South Carolina
Tennessee American Water Company	Chattanooga	Tennessee
Knox County First Utility District	Knoxville	Tennessee
Rivanna Water and Sewer Authority	Charlottesville	Virginia
Winchester Water Department	Winchester	Virginia
Aberdeen Water Department	Aberdeen	Washington
Salem Public Works Department	Salem	Washington

If the utility had no cement-mortar-lined pipe, no further questions were asked. If the utility knew of no cement-mortar lining corrosion problem, the question about system monitoring was asked to determine if the utility would be aware of a problem if it did exist. If the utility experienced cement-mortar lining corrosion problems, the survey was continued to obtain a thorough understanding of the problems.

Results

Thirty-one of the 33 surveyed utilities had distribution systems at least partly containing cement-mortar-lined pipes, either factory lined or lined in situ. Table 3.2 summarizes the survey results of these utilities.

Twenty-two utilities observed no cement-mortar lining corrosion problems. Most of them regularly monitor water quality throughout the distribution system and inspect removed pipe sections when possible.

Of the nine utilities experiencing problems with cement-mortar lining, seven reported water quality effects—that is, pH increases. Four of these utilities consider the problems to be significant; these four are discussed in detail in the

Table 3.2 Survey results—Utility satisfaction and problems with cement-mortar lining

Problem	Number of surveyed utilities	Comments
No observed problem	22	No pH change away from plant No customer complaints No structural deterioration
Significant water quality problem	4	pH increase to 9 or above Customer complaints No structural deterioration
Minor water quality problem	3	pH increase to 8–9 No customer complaints No structural deterioration
Structural problem	2	In one case, collapse of major section of lining in large pipe, caused primarily by physically abrasive water In the other case, pneumatic pressure used to reopen blocked service connections on newly lined pipes caused sections of lining surrounding the service pipe area to fall off
No cement-mortar-lined pipe used	2	
Total	33	

following section. The other three utilities noted that the pH increase is minor and not a concern.

Two utilities experienced structural problems with the lining. Their situations are presented later.

Significant Water Quality Problems

The four utilities with significant cement-mortar lining corrosion problems were (1) South Central Connecticut Regional Water Authority (South Central Connecticut RWA) in New Haven, CT; (2) Marin Municipal Water District (Marin MWD) in Corte Madera, CA; (3) Washington Suburban Sanitary Commission (Washington Suburban SC) in Laurel, MD; and (4) East Bay Municipal Utility District (East Bay MUD) in Oakland, CA. At each of these utilities, the typical cement-mortar lining corrosion problem is a significant pH increase (up to 9 or more) in low-flow areas, such as dead ends, particularly with newly lined pipe.

Table 3.3 presents finished water quality before contact with cement-mortar lining from these four utilities. Common features include pH between 7 and 8.5, low alkalinity, low hardness, and negative LI. (The LI is a measure of a water's tendency to dissolve or precipitate calcium carbonate. A negative LI denotes an aggressive water that tends to dissolve calcium carbonate. A water with a positive LI tends to precipitate calcium carbonate.) The South Central Connecticut RWA water was the most aggressive, with an LI of –2.2. Marin MWD had the highest hardness and alkalinity of the four and an LI closest to zero, –0.5. Water at Washington Suburban SC had an intermediate LI of –0.9. East Bay MUD (in the finished water from the plant associated with the cement-mortar problem) had an LI of –0.8, even with a pH of 8.4.

South Central Connecticut RWA

South Central Connecticut RWA's cement-mortar lining corrosion problem is the best-documented problem of the survey. In 1958, this utility began a program

Table 3.3 Finished water quality data for utilities with significant pH increases

Parameter	Marin MWD*	South Central Connecticut RWA†	Washington Suburban SC‡	East Bay MUD§
pH	8.0	7.2	7.8	8.4
Alkalinity as $CaCO_3$	64	16	32	21
Calcium as Ca	14	5	15	7
Hardness as $CaCO_3$	70	21	51	19
Silica as SiO_2	12	**	3	9
Iron as Fe	<0.1	<0.1	<0.1	<0.1
Zinc as Zn	0.6	<0.1	<0.1	<0.1
Phosphate as PO_4	0.8	0.7	<0.1	<0.1
LI	–0.5	–2.2	–0.9	–0.8

*San Geronimo Plant, 22 mgd (million gallons per day); 1988 averages.

†A regular sampling point in a main; June 1989 averages.

‡Patuxent Water Filtration Plant; 1988 averages.

§Lafayette Filter Plant; 1988 averages.

**Not reported.

of lining existing mains in situ. Initially, the lining program emphasized major transmission mains. Since 1980, however, the program has focused on sections of the distribution system with smaller-diameter mains.

The most significant cement-mortar lining corrosion problems occur near Milford, a community at the western edge of the system, where the low flow typical of system extremities is compounded by the presence of seasonal residences, which may be unoccupied and use no water during winter months. Customers in Milford have complained of skin irritation from the water. Utility personnel investigating the complaints found the water pH to be high, in the range of 9 to 10. The utility installed bleeders to purge the undesirable water and flush the pipes. In an area near Milford covering about 10 blocks, three bleeders run continuously 8 months of the year, and it can take 2 years after lining before the pH is controlled.

In continued investigations of cement-mortar lining corrosion, the utility found that waters from other recently lined pipes also had high pH, even mains in which the flow was relatively high. A month of normal flow, sometimes augmented with bleeders, usually stabilizes the pH at the normal filter plant effluent value of 7.2. In some high-flow situations, however, up to 2 years of purging is required.

South Central Connecticut RWA uses a zinc metaphosphate formulation for metallic corrosion control. The dose is 2 mg/L as Zn, which is applied in a solution of 50 percent P_2O_5 and 6.7 percent Zn.

Marin MWD

At the time of the survey, Marin MWD reported pH values from 10 to 12 in certain low-flow areas. The problems were attributed to the corrosion of cement-mortar lining in factory-lined pipe that, like most pipe at Marin MWD, was not seal coated. For over a year, the utility has combatted the high pH by flushing the lines with fresh water once per week for 20 to 30 minutes.

Marin MWD doses the finished water with one of two formulations of zinc orthophosphate for steel and copper corrosion control. One formulation has a $Zn:PO_4$ ratio of 1:3, by weight, and the utility controls the dose by maintaining a zinc concentration at 0.3 mg/L. The other formulation has a 1:1 $Zn:PO_4$ ratio, controlled at 0.5 mg/L Zn.

Washington Suburban SC

Washington Suburban SC's most recent cement-mortar corrosion problem occurred some time around 1986 in a rehabilitated pipe lined in situ. To combat a pH rise to 9, the utility flushed for about 2 months to stabilize the pH. The utility noted that continuous flushing was required to solve high pH problems. When periodic flushing was tried, the problem would recur the following day.

East Bay MUD

At the time of the survey, East Bay MUD had a cement-mortar corrosion problem with one residence in an unpopulated canyon. The residence, the only water user in the canyon in 1988, was connected with water service by factory-lined, seal-coated pipe. For over a year, the occupant was allowed to run an outdoor faucet continuously to flush high pH water from the lines. East Bay MUD reports that, ordinarily, one 10- to 15-minute flush is sufficient to combat high pH problems in newly installed factory-lined pipe.

Minor Water Quality Problems

The three surveyed utilities with minor cement-mortar corrosion problems were the Salem and Beverly (Massachusetts) Water Board, Greenville (South Carolina) Water System, and Boulder (Colorado) Water Department. The Salem and Beverly Water Board associated problems with in situ-lined pipe; the Greenville Water System and the Boulder Water Department associated problems with factory-lined seal-coated pipe. Each utility experienced pH increases to 8 to 9 in recently lined pipes, especially in low-flow areas. These problems were usually short-lived and resolved within months without flushing the lines.

Structural Problems

The San Francisco Water Department and the Boston Water and Sewer Commission experienced problems with the physical structure of cement-mortar lining. These structural problems are described below.

San Francisco Water Department

In 1978, the San Francisco Water Department suffered the collapse of thousands of feet of cement-mortar lining in a 96-in. (2.43-m) steel transmission line, which supplied the department with water from the Hetch Hetchy Reservoir in the Sierra Nevada Mountains. The pipe was factory-lined, with joints lined by hand (in situ). The low-alkalinity water was treated at the source with lime but no filtration, producing an LI of about –1.5 and a pH of 8.5. The water was physically abrasive, which was thought to be more significant than water chemistry in the collapse. Abrasive matter included particles in the raw water, lime grit, and sand washed from the transmission line's upstream rock tunnels.

The invert corroded first. Once it was weak, springline cracks occurred elsewhere, which were attacked by the abrasive water, now augmented by sand from eroded mortar, and whole sections of lining failed.

This lining collapse appears to be a site-specific problem not related to common utility circumstances. In addition to the abrasive water, an exacerbating factor may have been the shallow placement of the 96-in. (2.43-m) line, 12 ft (3.66 m) below the surface, where it could be affected by seasonal heating and cooling cycles.

The pipe has since been relined, with an extra 3-in. (7.6-cm) pad to protect the critical invert. Lime treatment is now based on achieving a target LI of zero downstream from the source. Furthermore, lime grit recovery is much improved. No lining deterioration problems have occurred since these steps were taken.

Boston Water and Sewer Commission

The Boston Water and Sewer Commission experienced problems when mains connected to residential service pipes were lined in situ. The cement mortar sealed the ⅝- and ¾-in. (1.6- and 1.9-cm) service openings. Pneumatic pressure, directed at 30 to 40 psi (pounds per square inch) (2.06 to 2.76 bar) from the service lines toward the lined pipe, was used to reopen the holes. In the process, an excess of cement mortar would fall off, about 2 in. (5.1 cm) around the holes. This problem was solved by back blowing at lower pressure, about 10 psi (0.69 bar).

Summary and Conclusions

All 31 surveyed utilities with cement-mortar-lined pipes produced water with low alkalinity, a parameter others have found to be important to lining corrosion. Seventy percent of these utilities observed no water quality or structural problems with cement-mortar-lined pipes. Nine utilities reported problems—significant, long-term pH increases; minor pH increases; and structural problems, where lining detached from pipe. The structural problems do not appear to be typical of industry experience. The lining collapse in the 96-in. (2.43-m steel) pipe is considered to be a site-specific incident related to physically abrasive water. The damage caused by back blowing sealed service openings was brought to an end by using lower air pressure. Therefore, the key cement-mortar lining corrosion problem was the pH increases to 9 or above, which required many months of flushing to stabilize.

The four utilities experiencing significant pH increases had aggressive, low-alkalinity waters, but with several differences in water quality. Alkalinities ranged from 15 to 65 mg/L, hardness from 20 to 70 mg/L, and LI from –0.5 to –2.2. This suggests that the corrosion problem involves more than water alkalinity. Perhaps other water quality parameters, water conditioning methods, cement-mortar composition, or lining procedures are important. Two of these utilities, South Central Connecticut RWA and Marin MWD, agreed to cooperate in a field test of various linings and water qualities to further investigate the cause(s) of corrosion in their distribution systems.

Chapter 4

Field Test Matrix and Procedures

Pilot-scale field tests were conducted at two water utilities to determine the conditions leading to cement-mortar lining corrosion. The participating utilities were South Central Connecticut RWA and Marin MWD, identified during the utility survey as sites experiencing problems with cement-mortar lining corrosion. The tests served the dual purpose of developing design recommendations to avoid cement-mortar lining corrosion, while helping the two participating utilities solve existing problems.

The concept of the field tests was to simulate low-flow (dead-end) conditions in an array of newly lined pipe sections carrying low-alkalinity water. This concept incorporated the major factors found in both the literature review and the utility survey to promote cement-mortar lining corrosion. The tests investigated how flow regime, lining method, seal coating, water quality, and mortar composition affected cement-mortar corrosion. This chapter describes the test apparatus, matrix, and procedures. Chapter 5 discusses test results.

Test Apparatus

Each field test operated with 10 lined pipe sections, mounted vertically on a support rig, as shown in Figures 4.1 through 4.3. Figure 4.1 is a conceptual diagram and Figures 4.2 and 4.3 are photographs of the actual rigs at South Central Connecticut RWA and Marin MWD, respectively.

The pipe sections were 3-ft (1-m) lengths of 4-in. (100-mm) nominal inside-diameter steel or ductile iron pipe (in situ-lined sections were steel, factory-lined sections were ductile iron). All but one section (section 6) were plumbed to receive plant finished water. Thus, these sections carried the same water that the utility's customers received. This water contained the utility's normal corrosion inhibitors. Section 6 was filled from a tank of water conditioned with different chemical additives to demonstrate the effect of different corrosion inhibitors. The water for section 6 was drawn upstream of the utility's corrosion inhibitor addition point, so that the water did not contain the utility's normal corrosion inhibitor. The section titled "Test Matrix," later in this chapter, describes the test conditions at each site.

All plumbing in the field test rigs—1-in. (25-mm) diameter tubing, valves, and flanges—was PVC (polyvinyl chloride) so that the feedwater contacted no metal, only inert PVC. This ensured that no metallic corrosion products from the

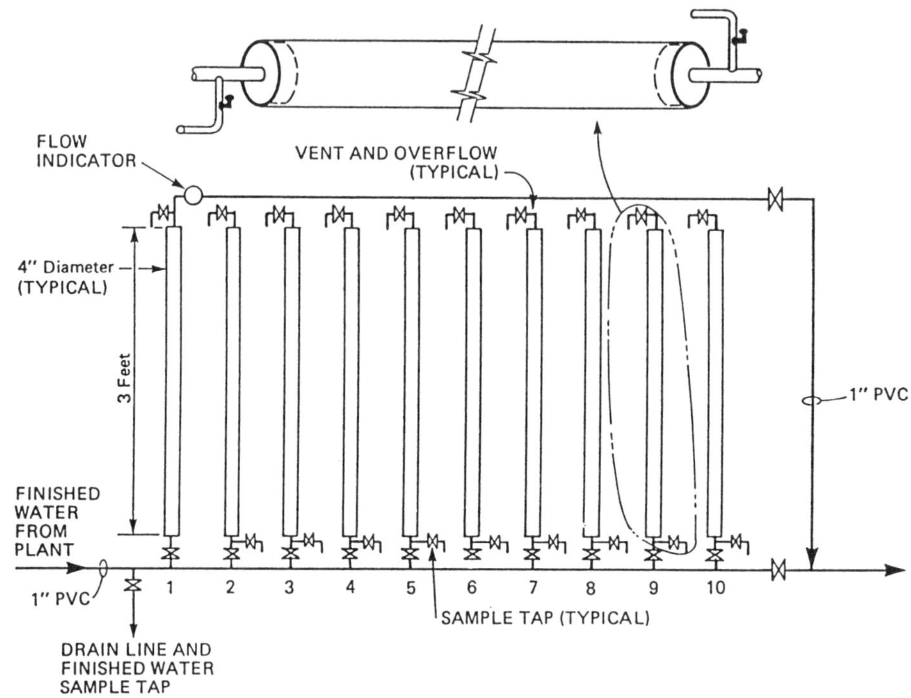

Figure 4.1 Conceptual diagram of pipe section array

Figure 4.2 South Central Connecticut RWA field test rig

Figure 4.3 Marin MWD field test rig

plumbing system could coat the cement mortar and confound analyses of test results. In addition, field test operators applied a waterproof epoxy (Porter International 7530, [Porter International, Louisville, KY], which is used in potable water systems) to seal any damaged areas of the pipe linings (refer to the description of the factory-lined pipes in the "Text Matrix" section). These precautions focused experimental results on just the interactions between water and cement-mortar linings.

The valving of all sections but section 6 gave operators the option of allowing continuous flow through the section or filling the section and then holding the water stagnant. However, only section 1 operated with continuous flow, at a rate of about 10 gpm (gallons per minute) [0.6 lps (liters per second)], measured by an in-line rotameter. This rate corresponds to a bulk velocity of about 0.26 fps (0.08 mps). The other sections were flushed (in other words, drained and refilled) either daily (section 2) or weekly (sections 2 through 10). When filled, these lined sections held about 1.6 gal (gallons) [6 L (liters)]. All sections were capped with tapped and plumbed PVC flanges, fitting directly against the edges of the lining.

Test Matrix

Table 4.1 presents the test matrix. All pipe sections were new pipe lined prior to the field tests. The experiments differed from each other in one parameter only, in order to isolate the effect of that parameter. The parameters were:

- Water flow regime.
- Lining method.
- Use of an asphaltic seal coat.
- Conditioning chemicals.
- Cement-to-sand ratio.
- Cement type.
- Use of a water-reducing admixture.
- Use of pozzolanic material.

Variations in the above parameters fell within the constraints of lining standards (AWWA 1989a; AWWA 1990).

All pipe sections were lined by the in situ drag trowel method, except sections 4 and 5, which were factory lined by the centrifugal method. In order to simulate in situ lining in the experiment, the drag trowel sections were lined above ground, using the same techniques normally applied underground. A professional pipe lining company, Heitkamp, Inc., of Watertown, CT, lined the drag trowel sections on April 20, 1990, according to ANSI/AWWA Standard C602-89. Pipe sections were stored on site, then shipped, filled with water, to the field test utilities. Shipment to South Central Connecticut RWA was by truck. Shipment to Marin MWD was by air freight. The McWane Cast Iron Pipe Company of Birmingham, AL, cut the factory-lined pipe sections from stock pipes, lined according to ANSI/AWWA Standard C104/A21.4-90.

The linings in sections 1, 2, 3, and 6 were identical, each lined by the in situ drag trowel method. These linings had the "reference" cement-mortar composition, defined as follows:

- A cement-to-sand ratio of 1:1, by volume.
- A slump of 6.5 in. (165 mm).*
- No asphaltic seal coat, admixture, or pozzolanic material.
- Type I portland cement, meeting ASTM Standard C150-89.

*Slump was used as a guide to the water content of the cement mortar. In typical lining operations, workers add water to the mortar until a desirable fluidity (slump) is achieved, producing a mortar that handles well when fresh, but which is also strong after it hardens. With too much slump, the mortar lacks strength and may not adhere to the pipe. With too little slump, the mortar is too difficult to pump. When preparing all the different in situ linings used in these field tests, the fabricator attempted to produce a slump of 4.5 to 5.5 in., with varying success.

Table 4.1 Test matrix

Pipe section number	Flow regime*	Slump† (in.)	Method of lining‡	Asphaltic seal coat	Chemical conditioning	Cement-to-sand ratio§	Portland cement type	Admixture?	Pozzolan?	Verbal description
1	Continuous	6 ½	In situ	No	Yes**	1.0:1.0	I	No	No	Reference mortar, continuous flow
2	Flush daily	6 ½	In situ	No	Yes**	1.0:1.0	I	No	No	Reference mortar, daily flush
3	Flush weekly	6 ½	In situ	No	Yes**	1.0:1.0	I	No	No	Reference mortar, weekly flush
4	Flush weekly	7 to 7 ½††	Factory	No	Yes**	~1.0:1.0††	II††	No	No	Factory lined without seal coat
5	Flush weekly	7 to 7 ½††	Factory	Yes	Yes**	~1.0:1.0††	II††	No	No	Factory lined with seal coat
6	Flush weekly	6 ½	In situ	No	Yes‡‡	1.0:1.0	I	No	No	Reference mortar, alternative water conditioning
7	Flush weekly	4 ¾	In situ	No	Yes**	1.0:1.5	I	No	No	High sand
8	Flush weekly	4 ¾	In situ	No	Yes**	1.0:1.0	II	No	No	Type II cement
9	Flush weekly	3 ½	In situ	No	Yes**	1.0:1.0	I	Yes§§	No	Water-reducing admixture
10	Flush weekly	5 ½	In situ	No	Yes**	§§	I	No	Yes***	Pozzolan

*Section 1 had continuous flow at about 10 gpm. Section 2 was stagnant, with water changed daily. Sections 3 through 10 were stagnant, with water changed weekly.

†Slump is a measure of mortar fluidity. The slump test is described in ASTM C143.

‡In situ lining was drag trowel method. Factory lining was centrifugal method.

§Volume ratio.

**The chemical conditioner for South Central Connecticut RWA sections 1–5 and 7–10 was the Shannon Chemical Corporation's (Malvern, PA) Shan-No-Corr. It is a zinc metaphosphate formulation (1 part Zn to 7.5 parts metaphosphate, as P_2O_5; total dose = 2 mg/L). The chemical conditioner for Marin MWD sections 1–5 and 7–10 was Technical Products Corporation's, (Portsmouth, VA) V939. It is a 1:3 zinc orthophosphate formulation and was added to provide a zinc dose of 0.3 mg/L.

††Factory-lined pipe came from stock. Vendor used Type II cement. Exact cement-to-sand ratios and slump data were not known for these particular specimens. The table presents typical values, provided by the vendor.

‡‡The alternative chemical conditioner for South Central Connecticut RWA section 6 was Technical Products Corporation's V939. It is a 1:3 zinc orthophosphate formulation. The chemical was added to provide a zinc dose of 0.3 mg/L. The alternative chemical conditioner for Marin MWD section 6 was a Stiles Chem Product (Stiles Chemical Company, Zion, IL). It is a combination polyphosphate–orthophosphate. It was added to provide a dose of 0.8 mg/L, as total phosphate. Note that the water for each section 6 was drawn upstream of the treatment plant's corrosion inhibitor addition point.

§§Used 16 fluid ounces of Sikament 300 (Sika Corporation, Lyndhurst, NJ) per cubic foot (94 pounds) of cement. Sikament 300 is a Type F high-range water-reducing admixture, nonair-entraining, complying with requirements of ASTM C494, Type F.

***Used pozzolanic fly ash as a partial substitute for cement. Substituted to produce a fly ash-to-cement-to-sand volume ratio of 0.2:0.8:1.0.

For sections 1, 2, and 3, the variable parameter was the inlet water flow regime. Section 1 carried continuously flowing water. Section 2 held stagnant water, flushed daily. Section 3 also held stagnant water, but the pipe section was flushed weekly. Weekly flushing was expected to simulate dead ends most closely, with daily flushing representing a less severe low-flow situation. The section with continuous flow simulated cement-mortar corrosion in transmission mains, where lining impacts might still occur even if constant flushing masks pH or other water quality effects.

All remaining pipe sections were flushed weekly, following the flow regime of section 3. These sections tested the effects of various factors—section 4, lining method; section 5, seal coat; section 6, conditioning chemicals; and sections 7 through 10, mortar composition. The following list describes sections 4 through 10:

- Section 4 was cut from factory-lined pipe, lined by the centrifugal method with a mortar containing Type II cement. It did not have an asphaltic seal coat, nor did the mortar contain a water-reducing admixture or pozzolanic material. Because the pipe vendor does not keep data for individual pipe sections, it was not possible to determine the cement-to-sand ratio or slump of the stock mortar for either section 4 or section 5. The values in Table 4.1 are typical of operations at the factory that lined these pipe sections.

- Section 5 was the same as section 4, except that it had a factory-applied asphaltic seal coat. (This is the standard way factory-lined ductile iron pipe is delivered.) Because cutting sections 4 and 5 damaged the linings slightly at the ends of the sections, utility personnel at the field test sites coated the cut ends with epoxy before start-up.

- Section 6 used the reference mortar, but contained a water that was conditioned differently from the normal plant finished water. At the South Central Connecticut RWA site, the alternative conditioner was a 1:3 (weight ratio) zinc orthophosphate formulation, added at 0.3 mg/L as Zn. This chemical was Technical Products V939, manufactured by the Technical Products Corporation of Portsmouth, VA. At the Marin MWD site, the alternative conditioner was a combination polyphosphate–orthophosphate, dosed at 0.8 mg/L as total PO_4. This chemical was a Stiles chemical product, manufactured by the Stiles Chemical Company at Zion, IL. The utilities prepared the conditioned waters in quantities large enough to serve the field tests in one or two batches.

  The water for each section 6 was drawn upstream of the utility's corrosion inhibitor addition point, so that the water did not carry the normal conditioning agents. The water to all other test sections carried the utility's normal corrosion inhibitors. Table 4.1 describes these inhibitors.

- Section 7 had the reference mortar composition, except for a higher proportion of sand in the mixture. The reference mortar had the highest cement-to-sand ratio allowed by the ANSI/AWWA C602 standard, 1:1, by volume; section 7 had the lowest, 1:1.5, by volume. The slump was 4.75 in. (12 mm).

- Section 8 had the reference mortar composition, except for the use of Type II portland cement, a low tricalcium aluminate cement, which is normally used to provide protection against sulfate attack. The Type II cement used in the field tests satisfied ASTM C150-89. The slump of section 8 mortar was 4.75 in. (120 mm).
- Section 9 used a water-reducing admixture—Sikament 300, an ASTM C494 Type F, high-range, nonair-entraining admixture. This admixture, which is manufactured by the Sika Corporation of Lyndhurst, NJ, improves mortar fluidity, thus allowing the desired slump to be achieved with less water. Less water, in turn, improves mortar strength and decreases its porosity. Section 9 slump was 3.5 in. (90 mm).
- Section 10 substituted an ASTM C618 Class F pozzolanic material (fly ash) for part of the cement, creating a pozzolan-to-cement-to-sand ratio of 0.2:0.8:1.0, by volume (assuming the fly ash had the same bulk density as the cement, 94 lb/ft^3 (pounds per cubic foot) (1506 kg/m^3 [kilograms per cubic meter]). Its slump was 5.5 in. (140 mm). Because pozzolans react with lime to form cementitious compounds, section 10 investigated if the pozzolanic mortar would reduce lime leaching.

Table 4.2 presents the characteristics of the hardened cement mortars used for in situ lining—compressive strength (ASTM C39-86), flexural strength (ASTM C78-84), air content (ASTM C173-78), and water absorption (ASTM C642-90). Professional Service Industries, Inc., of Bloomfield, CT, performed the hardened mortar strength, air content, and water absorption measurements.

Five cement-mortar types were used to line the eight in situ test sections. A review of Table 4.2 shows the following:

- The reference mortar (sections 1, 2, 3, and 6) was strong compared with most of the other sections. The compressive strength on the 28th day after curing was 8,280 psi (571 bar). The flexural strength was the highest of the sections tested, for both failure load (15,500 lbf [pounds

Table 4.2 Characteristics of hardened cement mortars used for in situ lining

		Compressive strength (psi)		Flexural strength after 28 days			
Mortar type	Used in pipe section	After 7 days	After 28 days	Failure load (lbf)	Modulus of rupture (psi)	Air content (%)	Water absorption (%)
Reference	1,2,3,6	6,220	8,280	15,500	1,260	4.8	9.8
High sand	7	4,000	4,990	*	*	9.4	9.5
Type II cement	8	6,970	7,980	*	*	4.1	7.8
Admixture	9	7,500	9,050	13,500	1,105	7.2	7.0
Pozzolan	10	5,800	7,300	12,550	1,020	4.0	8.3

*Test not done.

force]) (68,944 N [Newtons]) and the modulus of rupture (1,260 psi [87 bar]). The reference mortar had a moderate air content of 4.8 percent and a high water absorption of 9.8 percent.

- The high sand mortar (section 7) had the lowest compressive strength (4,990 psi [344 bar]), a value that increased only 14 percent between the 7th and 28th days after curing—the smallest increase of all the mortar types. Flexural strength was not tested for the high sand mortar. The air content and water absorption were high at 9.4 and 9.5 percent, respectively.
- The Type II cement mortar (section 8) had a compressive strength comparable to that of the reference mortar, nearly 8,000 psi (552 bar) after 28 days. Flexural strength was not tested. It had a low air content of 4.1 percent and a moderate water absorption of 7.8 percent.
- The admixture mortar (section 9) had the highest compressive strength (9,050 psi [624 bar] after 28 days) and a flexural strength approaching 90 percent of that of the reference mortar. The moderate air content was 7.2 percent and the water absorption 7.0 percent.
- The pozzolan mortar (section 10) had a moderate compressive strength (7,300 psi [503 bar] after 28 days) and the lowest flexural strength of the sections tested—a failure load of 12,550 lbf (55,822 N) and a modulus of rupture of 1,020 psi (70 bar). It had a low air content of 4.0 percent and a moderate water absorption of 8.3 percent.

There is no corresponding information for the factory-lined sections, because the pipe vendor does not keep such data for individual shipments.

Operating Procedures

Operators from the two participating utilities conducted the field tests under the direction of the research team. Tests began on April 30, 1990, at South Central Connecticut RWA and on May 1, 1990, at Marin MWD. The tests ran for 3 months (13 weeks).

Operators started up the tests by initiating a flow of plant finished water through section 1. Next, they filled the remaining sections (except section 6) with plant finished water and closed the valves to hold the water stagnant in the sections. Finally, they filled section 6 with water conditioned with the alternative corrosion inhibitor and held the water stagnant.

Operating procedures involved inspecting the system for leaks or other problems, reading the flow rate through section 1, sampling, and flushing pipe sections. The paragraphs below further describe sampling and analysis.

Sampling

Operators collected all samples in plastic buckets. The plant finished water was sampled from a tap preceding distribution to the rig. Section 1 was sampled at its overhead discharge. All remaining sections were sampled during the flushing operation, as follows:

1. A few pints of water were drained from the bottom of the section and used to purge the sample tap and rinse the bucket, after which this water was discarded.

2. The rest of the water was then drained into the sample bucket for analysis. Thus, the sample consisted of nearly the entire contents of the section, ensuring representative results.

Operators measured temperature and pH immediately. Then they poured the well-mixed contents of the bucket into sample bottles and delivered the samples to the utility analytical laboratory.

After sampling, operators refilled the sections with fresh water. They used only enough water to fill the sections completely, avoiding extensive flushing by closing the inlet valve as soon as water exited the upper discharge pipe.

Analysis

The samples were analyzed frequently for temperature, pH, alkalinity, and calcium, with less frequent monitoring of silica, zinc, magnesium, phosphate, iron, and manganese. The analytical schedule began with intense monitoring for the first 2 weeks of operation, when water quality changes were the greatest, then tapered off as the remaining weeks exhibited less change.

First 2 Weeks

Table 4.3 presents the sampling and analytical schedules for the first 2 weeks. Operators sampled plant finished water (i.e., inlet water) and the continuous discharge from section 1 daily.

Table 4.3 Sampling and analysis during first two weeks

Sample point	Temperature °C	pH	Alk. (mg/L $CaCO_3$)	Ca (mg/L $CaCO_3$)	SiO_2, Zn, Mg, ortho PO_4, total PO_4, Fe, Mn (mg/L as ion)
Inlet water	Daily	Daily	MWF*	MWF	Weekly†
Section 1‡	Daily	Daily	MWF	MWF	Weekly
Section 2	Weekly	Weekly	Weekly	Weekly	Weekly
Section 3	Weekly	Weekly	Weekly	Weekly	Weekly
Section 4	Weekly	Weekly	Weekly	Weekly	Weekly
Section 5	Weekly	Weekly	Weekly	Weekly	Weekly
Section 6	Weekly	Weekly	Weekly	Weekly	Weekly
Section 7	Weekly	Weekly	Weekly	Weekly	Weekly
Section 8	Weekly	Weekly	Weekly	Weekly	Weekly
Section 9	Weekly	Weekly	Weekly	Weekly	Weekly
Section 10	Weekly	Weekly	Weekly	Weekly	Weekly
Conditioned water	Weekly	Weekly	Weekly	Weekly	Weekly

*MWF means Monday, Wednesday, Friday.

†Weekly samples taken on Monday.

‡Section 1 flow rate (gpm) recorded daily.

Sections 2 through 10 were sampled when flushed, daily for section 2 and weekly (on Mondays) for the rest. The conditioned water (the water used to charge section 6) was sampled weekly.

Temperature and pH were measured every day a sample was taken. Alkalinity and calcium were measured three days per week (Monday, Wednesday, and Friday) for the inlet water and sections sampled daily, and weekly for the other sections. The remaining parameters were measured weekly for all sections, the inlet water, and the supply of alternatively conditioned water.

Next 11 Weeks

The analytical schedule relaxed after the first 2 weeks, as shown in Table 4.4. Operators continued to measure the temperature and pH of the inlet water and the discharges from sections 1 and 2 on a daily basis, whereas, alkalinity and calcium were measured only once per week (on Mondays). For all other sections and the alternatively conditioned water, operators measured temperature, pH, alkalinity, and calcium once per week, during flushing. The analyses for silica, zinc, magnesium, phosphate, iron, and manganese were conducted three times during the final 11 weeks of the test—on May 28, June 25, and July 30, 1990.

Marin MWD also conducted additional analyses for THMs (trihalomethanes), PAHs (polynuclear aromatic hydrocarbons), and volatile organic compounds. Trihalomethanes—disinfection (usually chlorination) by-products that may cause health problems—form more readily as pH increases (Stevens et al. 1989). Thus, THM concentrations in section 1 discharge (which had THM concentrations expected to be similar to THM concentrations in the inlet water) were compared with concentrations in water from sections 4 and 5, which had elevated pH values, to determine if cement-mortar lining corrosion or organics leached from the seal coat might contribute to THM formation. Polynuclear aromatic hydrocarbons are the priority pollutants most likely to be found in asphaltic seal coats (Miller et al. 1982). Thus, section 1 discharge PAH concentrations were compared with

Table 4.4 Sampling and analysis during final 11 weeks

Sample point	Temperature °C	pH	Alk. (mg/L $CaCO_3$)	Ca (mg/L $CaCO_3$)	SiO_2, Zn, Mg, ortho PO_4, total PO_4, Fe, Mn (mg/L as ion)
Inlet water	Daily	Daily	Weekly	Weekly	Three times*
Section 1†	Daily	Daily	Weekly	Weekly	Three times
Section 2	Weekly	Weekly	Weekly	Weekly	Three times
Section 3	Weekly	Weekly	Weekly	Weekly	Three times
Section 4	Weekly	Weekly	Weekly	Weekly	Three times
Section 5	Weekly	Weekly	Weekly	Weekly	Three times
Section 6	Weekly	Weekly	Weekly	Weekly	Three times
Section 7	Weekly	Weekly	Weekly	Weekly	Three times
Section 8	Weekly	Weekly	Weekly	Weekly	Three times
Section 9	Weekly	Weekly	Weekly	Weekly	Three times
Section 10	Weekly	Weekly	Weekly	Weekly	Three times
Conditioned water	Weekly	Weekly	Weekly	Weekly	Three times

*The three dates were May 28, June 25, and July 30, 1990.

†Section 1 flow rate (gpm) recorded daily.

concentrations in water from section 5 to determine if the seal coat leached PAHs. Marin MWD also measured PAH in water from section 4, as a check of whether cement-mortar corrosion is related to PAH concentrations. The analyses for volatile organic compounds provided additional information about organics leaching.

Post-Test Lining Inspections

At the completion of the field tests, operators dismantled the test rigs, leaving the pipe sections filled with water. The sections were transported to Calcoast Analytical–ITL (a materials testing laboratory in Emeryville, CA), where the linings were examined for visual appearance, loose sand, hardness, calcium loss with depth, and the composition of scales (if any). Also examined were "control" pipe sections, which had linings identical to those in the field tests. The control pipes had been set aside at the beginning of the test and were not subjected to any flow regime. The control sections had been filled with water to keep the linings from drying out.

In addition, South Central Connecticut RWA dug up three 1-ft (0.3-m) sections of in situ-lined pipe that had been in municipal water distribution service for 1, 6, and 12 years. Analyzing these sections provided information about the effects of long-term exposure to water on cement mortar.

Appearance, Loose Sand, and Hardness

After draining the sections and allowing them to dry, researchers visually inspected each lining for color, texture, defects, and the presence of scale deposits. Researchers sought evidence of loose sand both visually and by brushing the lining with a wire brush. Hardness was evaluated qualitatively by scratching the lining with a thumbnail.

Scale Composition, Calcium Loss, and pH Effects

The materials testing laboratory developed and operated a system to scrape thin layers, up to 150 mils (3.80 in.), from the lining and collect the shavings. The system consisted of three abrasive pads, 3-in. × 0.5-in. (7.5-mm × 1.3-mm), pressed by spring tension against the lining and rotated by a motor. A vacuum tube collected the shavings onto glass wool, from which they were subsequently removed for analyses. Calipers gauged the depths of the cut layers. This method could scrape off a thin scale deposit, then continue to cut a series of mortar layers for calcium analysis. The layer-by-layer calcium analyses provided profiles of calcium content versus mortar depth. Profiles of pH were also developed.

Scales were analyzed by dispersive X-ray fluorescence, yielding concentrations of calcium, silica, zinc, magnesium, phosphate, iron, and manganese. Because the method is not effective for elements with atomic numbers less than 10, carbon was determined by igniting the dry sample in a muffle furnace at 550°C (Celsius) (1022°F [Fahrenheit]) and measuring the weight loss. Assuming any weight loss was caused by the release of carbon dioxide, the original carbon concentration of the scale was calculated.

Table 4.5 Field test sections analyzed for scale composition, calcium loss, and pH effects

		Test performed for:		
Section	Lining description	Control	South Central Connecticut RWA	Marin MWD
1	Reference, continuous flow	Calcium	Scale/calcium/pH	Scale/calcium
2	Reference, daily flush	*	Calcium/pH	Calcium
3	Reference, weekly flush	*	Calcium/pH	Scale/calcium
4	Factory, no seal coat	Calcium	NT†	Calcium
5	Factory, seal coat	NT	NT	NT
6	Reference, conditioned water	*	Calcium/scale	Scale
7	High sand	NT	NT	Scale
8	Type II cement	NT	NT	Calcium
9	Admixture	Calcium	NT	NT
10	Pozzolan	NT	NT	NT

*One pipe section represented all control linings for sections 1, 2, 3, and 6 (reference mortars).

†NT means not tested.

The calcium content of the lining shavings was measured by the California Department of Transportation's Test 403, a titration procedure that involves preliminary sample digestion steps (State of California 1983). Calcium concentration results are expressed in units of percent CaO in the solid sample.

A portion of the shavings produced for calcium analysis were slurried with deionized water for pH analysis. The slurry was allowed to settle before measurement with a pH probe.

Table 4.5 shows which field test sections had linings analyzed for scale composition, calcium loss, or pH effects. Scales were analyzed for those sections that had visible scale deposits. In a few instances, the first few mils of surface were analyzed for scalants even if no scales were visible. The reason for these analyses was an indication (via water analyses) that scalants might have adsorbed or precipitated on the mortar surface. A key goal of the calcium tests was to evaluate the impact of flow regime on lime leaching. Other calcium tests were run on sections that had notable pH, alkalinity, or calcium results. Each aged pipe section dug up by South Central Connecticut RWA was analyzed for scale composition, calcium loss, and pH.

Chapter 5

Field Test Results

The objective of the field tests was to identify the factors that have the most influence on the corrosion of cement-mortar lining. Field tests at two sites investigated the following factors:

- Water flow regime through lined pipe sections.
- Method of lining—drag trowel (in situ method) and centrifugal (factory method).
- Use of a seal coat (for centrifugally lined pipe only).
- Use of water-conditioning chemicals.
- Cement-mortar composition.

The test matrix was set up so that each test section differed from the others in one parameter only. This was done to isolate the effect of that parameter (see Table 4.1). Cement-mortar corrosion was monitored in two ways:

- Effect on quality of water in the pipe sections.
- Effect on lining structure.

The two field test sites were South Central Connecticut RWA, in New Haven, CT, and Marin MWD, in Corte Madera, CA.

This chapter presents the field test results, first discussing water quality effects, then lining structure effects. The chapter closes by summarizing key lessons learned at both sites.

Water Quality Effects

The most important water quality parameters were pH, alkalinity, and calcium. These parameters relate to lime leached from the cement mortar. These values changed (increased) significantly upon exposure of water to the cement-mortar linings. Also monitored were silica, zinc, magnesium, phosphate, iron, and manganese. Increases in silica content upon exposure indicate leaching of cement compounds. Literature sources identified the remaining parameters as having the potential to retard cement-mortar corrosion by forming precipitates on or within the lining surface. In addition, the presence of iron could indicate that water in a pipe section contacted unintentionally exposed metal.

Marin MWD also conducted limited analyses for THMs, PAHs, and volatile organic compounds. The section on Marin MWD water quality results discusses these organic parameters.

The following discussions of water quality results are accompanied by summary bar charts of pH, alkalinity, and calcium data. These charts present data from six selected weeks throughout the test period, weeks 1, 2, 3, 6, 9, and 13 (the final week). This selection of data represents trends accurately without cluttering the presentation. Appendix A contains the complete data, as well as similar bar charts for other inorganic parameters.

There are two series of bar charts for each field test site. The "A" series, always the top chart in a figure, presents a parameter's results for the following sections:

- Inlet water.
- Section 1—Reference mortar,* continuous flow.
- Section 2—Reference mortar, daily flush.
- Section 3—Reference mortar, weekly flush.
- Section 4—Centrifugally lined, without seal coat, weekly flush.
- Section 5—Centrifugally lined, with seal coat, weekly flush.
- Section 6—Reference mortar, alternative conditioning chemicals, weekly flush.

Thus, the "A" series relates to the effects of the first four corrosion factors investigated—flow regime, lining method, seal coat, and conditioning chemicals.

The "B" series concerns the effects of variations in cement-mortar composition alone. Always the bottom chart in a figure, the "B" series presents a parameter's results for the sections below:

- Inlet water.
- Section 3—Reference mortar, weekly flush.
- Section 7—High sand-to-cement ratio, weekly flush.
- Section 8—Type II cement, weekly flush.
- Section 9—Water-reducing admixture, weekly flush.
- Section 10—Pozzolan, weekly flush.

The inlet water is on both series of charts to provide baseline water quality values from before the water contacted the linings. The charts show the inlet water values as a bold horizontal line, wide enough to reflect normal variations. (Because the results from section 1—reference mortar with continuous flow—always matched inlet water results, the same bold line represents section 1 results as well). Also, the reference mortar section, flushed weekly (section 3), is on both series of

---

*The "reference mortar" had the standard composition in terms of cement and sand proportions, cement type, and additives. Table 4.1 describes the reference mortar in detail and how other mortars differed from it.

charts, serving as the standard against which most other weekly flushed sections are compared. The times when section 3 is not the standard for comparison concern the two centrifugally lined (factory-lined) sections. When evaluating the effect of fabrication technique, compare the performance of the centrifugally lined section without a seal coat (section 4) against performance of drag trowel-lined section 8. These two sections have common flow regimes and are both fabricated of Type II cement. To evaluate the effect of a seal coat, compare the results of sections 4 and 5. Except for the seal coat, these two sections have identically fabricated linings and identical flow regimes.

South Central Connecticut RWA Water Quality Results

This section describes the inlet water quality, followed by discussions of how the corrosion factors—flow regime, lining method, seal coating, conditioning chemicals, and mortar composition—affected pH, alkalinity, and calcium. Table 5.1 summarizes the results for these three key parameters for all 10 pipe sections, as well as the inlet water and the conditioned water. The section concludes with a discussion of the responses of silica and the other monitored inorganic parameters.

Inlet Water

The top row of Table 5.2 presents average values and standard deviations for pH, alkalinity, and calcium in the inlet water during the test.

The inlet water at South Central Connecticut RWA was aggressive, with an LI of –2.5 (slightly lower than the historical value presented in Table 3.3). The small standard deviations show that the water quality was very steady at this field test site.

Flow Regime

The three flow regimes were continuous flow, water held stagnant with a daily flush, and water held stagnant with a weekly flush. A "flush" means that a section was emptied, then immediately refilled; there was no flow otherwise. Each of the pipe sections used to evaluate flow regime (sections 1, 2, and 3, respectively) was lined with the reference cement mortar by the drag trowel method.

The 10-gpm (0.63-lps) flow rate through the continuous-flow section produced a hydraulic detention time of about 10 seconds. This brief contact did not change the water quality. Water from the continuous-flow section had essentially the same pH, alkalinity, and calcium values as the inlet water, as shown in Table 5.2.

Water held in the pipe sections for 1 day or 1 week between flushes leached significant amounts of lime from the lining, as indicated by increased pH, alkalinity, and calcium concentrations.

pH. Figure 5.1A demonstrates the effect of flow regime on pH. The continuous-flow pH values, as explained above, are the same as the inlet water values shown in the figure. The two left-hand bars in each group are pH values from the daily flushed and weekly flushed sections.

Continuous-flow pH values did not increase over inlet values, averaging 7.2 throughout the 13-week test. Daily flushed pH values were initially 4 to 5 units higher than the inlet values, at 11 to 12, then decreased to near 10.0. The weekly flushed pH was 12.5 after the first week, more than 5 units higher than the inlet value. Weekly flushed pH values decreased to near 11.0 by the test end, remaining

Table 5.1 Summary of South Central Connecticut RWA field test results

Section	pH	Alkalinity (mg/L $CaCO_3$)	Calcium (mg/L $CaCO_3$)
Inlet water (fed to all sections but section 6)	Steady at 7.2	Steady at 14	Steady at 13
Alternatively conditioned (fed to section 6 only)	Steady at 6.8	Steady at 10	Steady at 13
1. Reference, continuous flow	Steady at 7.2	Steady at 14	Steady at 13
2. Reference, daily flush	11.7 after first day, daily pH values jumped around, varied from 9.9 to 10.4 for final 3 weeks of test	390 after first day, stabilized at 40 after 3 weeks	80 after first day, stabilized at 25 to 30 after 3 weeks
3. Reference, weekly flush	12.5 after first week, decreased to 10.9 to 11.0, may not have stabilized during test	760 after first week, stabilized at 70 after 9 weeks	135 after first week, stabilized at 35 after 7 weeks
4. Centrifugally lined, no seal coat, weekly flush	11.9 after first week, decreased to 11.1 to 11.3, may not have stabilized during test	280 after first week, decreased to 80, still decreasing at end of test	30 after first week, increased over next 4 weeks to 150, then fell gradually to 110, on final week, calcium was 80
5. Centrifugally lined, seal coat, weekly flush	11.8 after first week, stabilized at 9.8 to 10.0 after 9 weeks	200 after first week, stabilized at 30 after 9 weeks	30 after first week, stabilized at 15 to 20 after 3 weeks
6. Reference, alternative conditioning chemicals, weekly flush	12.6 after first week, stabilized at 11.3 to 11.4 after 11 weeks	920 after first week, decreased to 110, still decreasing at end of test	290 after first week, varied from 100 to 130 for final 8 weeks
7. High sand-to-cement ratio, weekly flush	12.5 after first week, stabilized at 11.1 to 11.3 after 10 weeks	690 after first week, decreased to 85, still decreasing at end of test	140 after first week, 210 after second week, stabilized at 80 to 90 after 6 weeks
8. Type II cement, weekly flush	12.4 after first week, stabilized at 11.3 after 11 weeks	680 after first week, decreased to 110, still decreasing at end of test	60 after first week, 195 after second week, 215 after third week, decreased to 85, still decreasing at end of test
9. Water-reducing admixture, weekly flush	12.5 after first week, stabilized at 11.2 after 11 weeks	740 after first week, decreased to 85, still decreasing at end of test	55 after first week, 180 after second week, decreased to 60, still decreasing at end of test
10. Pozzolan, weekly flush	12.6 after first week, stabilized at 11.1 after 11 weeks	790 after first week, stabilized at 80 after 10 weeks	80 after first week, 150 after second week, stabilized at 50 to 60 after 10 weeks

Table 5.2 South Central Connecticut RWA inlet versus continuous-flow water quality

Water description	pH		Alkalinity ($mg/L\ CaCO_3$)		Calcium ($mg/L\ CaCO_3$)	
	Average	Standard deviation	Average	Standard deviation	Average	Standard deviation
Inlet water	7.19	0.09	14	1.2	13	1.6
Continuous flow (section 1 water)	7.18	0.09	14	1.2	13	1.5

1 or more pH units higher than daily flushed values. The continued decreasing trend indicates that the weekly flushed water may not have stabilized with respect to pH in 13 weeks. ("Stabilized" is defined within the constraints of the 13-week field test. It means the results were stable for 3 or more weeks.)

Alkalinity. The effect of flow regime on alkalinity is shown in Figure 5.2A. Again, the continuous-flow alkalinity did not increase relative to the inlet water concentration of 14 mg/L $CaCO_3$, but the daily and weekly flushed results show large alkalinity increases relative to the inlet water. The daily flushed alkalinity was 390 mg/L $CaCO_3$ after the first day, decreasing rapidly to a stable 40 mg/L $CaCO_3$ by the end of the test. (Refer to appendix A for the first days' values. The charts show only weekly values.) The weekly flushed alkalinity was 760 mg/L $CaCO_3$ after the first week, stabilizing at 70 mg/L $CaCO_3$ after 9 weeks.

Calcium. Figure 5.3A shows how flow regime affected calcium concentrations. The continuous-flow water remained at the inlet concentration of 13 mg/L $CaCO_3$. The daily flushed concentration was 80 mg/L $CaCO_3$ after the first day, but quickly stabilized at 25 to 30 mg/L $CaCO_3$ after 3 weeks. The first week's calcium concentration in the weekly flushed section was 135 mg/L $CaCO_3$, stabilizing at 35 mg/L $CaCO_3$ after 7 weeks.

Summary. The progression of values from continuous flow to daily flush to weekly flush clearly shows how pH, alkalinity, and calcium increased the longer water remained in the section. Even after 13 weeks of daily flushing, contact with the lining produced high-pH waters, with alkalinity and calcium concentrations substantially higher than the inlet water. Weekly flushed impacts were greater. Alkalinity concentrations in the first flushed waters were extremely high, 390 and 760 mg/L $CaCO_3$ in the daily and weekly flushed sections, respectively. While water flowing continuously past the lining did not exhibit increased levels of dissolved lime compounds, this does not mean that leaching did not occur in this section. However, the continuous flushing kept any leaching from noticeably affecting water quality because of dilution.

Drag Trowel Versus Centrifugal Lining

Section 4 was lined by the centrifugal method with a Type II cement mortar. The best way to determine the effect of this lining method is to compare the

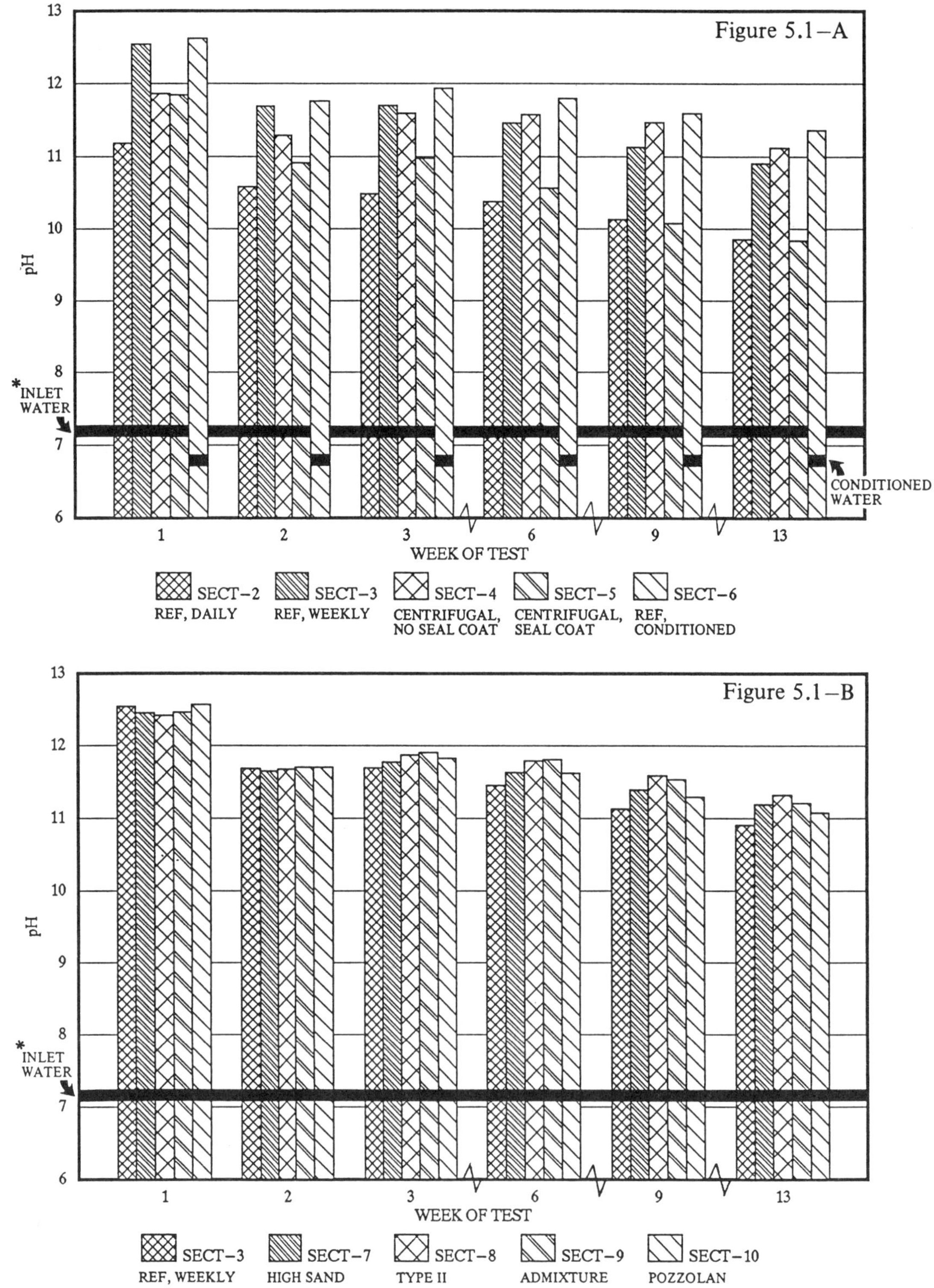

*SECT–1 (Reference Mortar, Continuous Flow) had the same composition as the Inlet Water.

Figure 5.1 South Central Connecticut RWA pH results

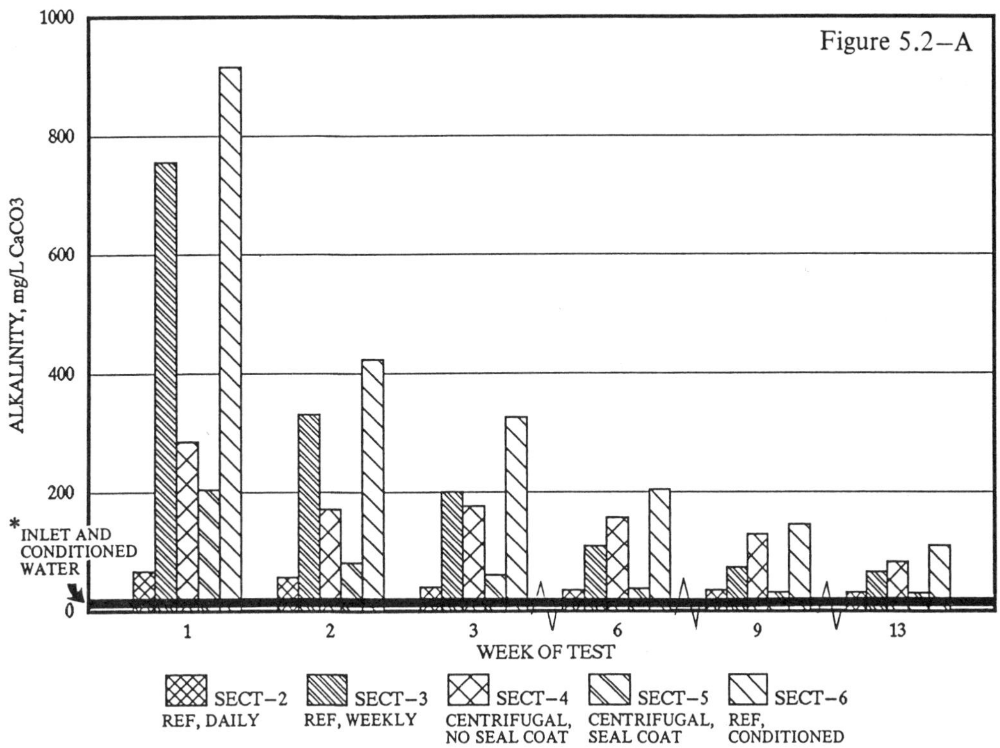

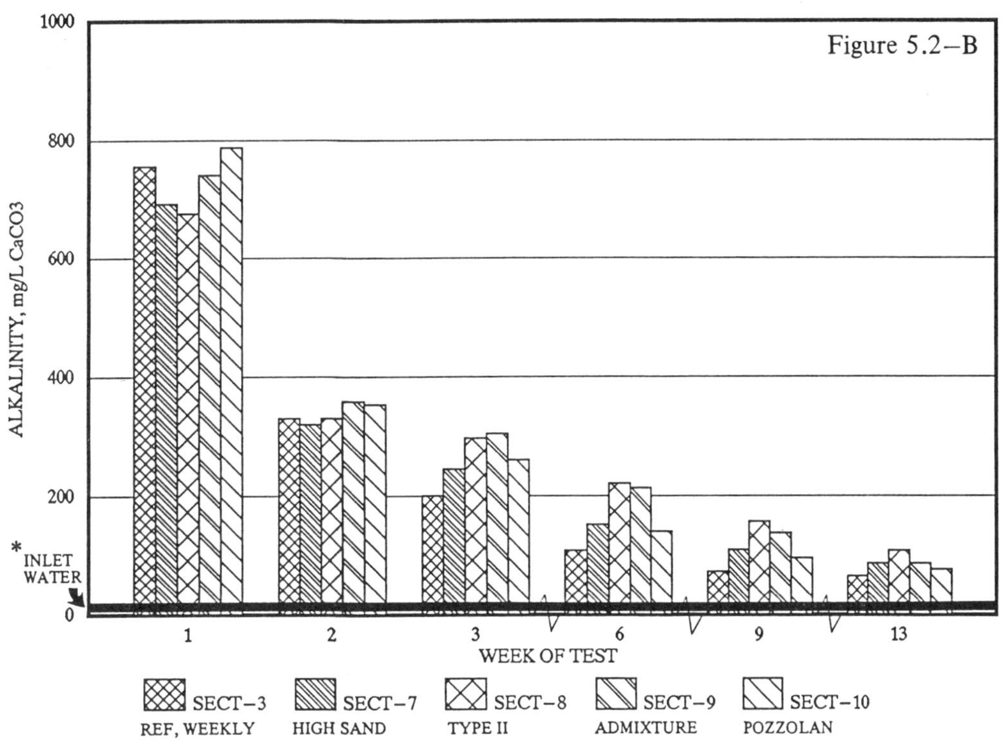

*SECT–1 (Reference Mortar, Continuous Flow) had the same composition as the Inlet Water.

Figure 5.2 South Central Connecticut RWA alkalinity results

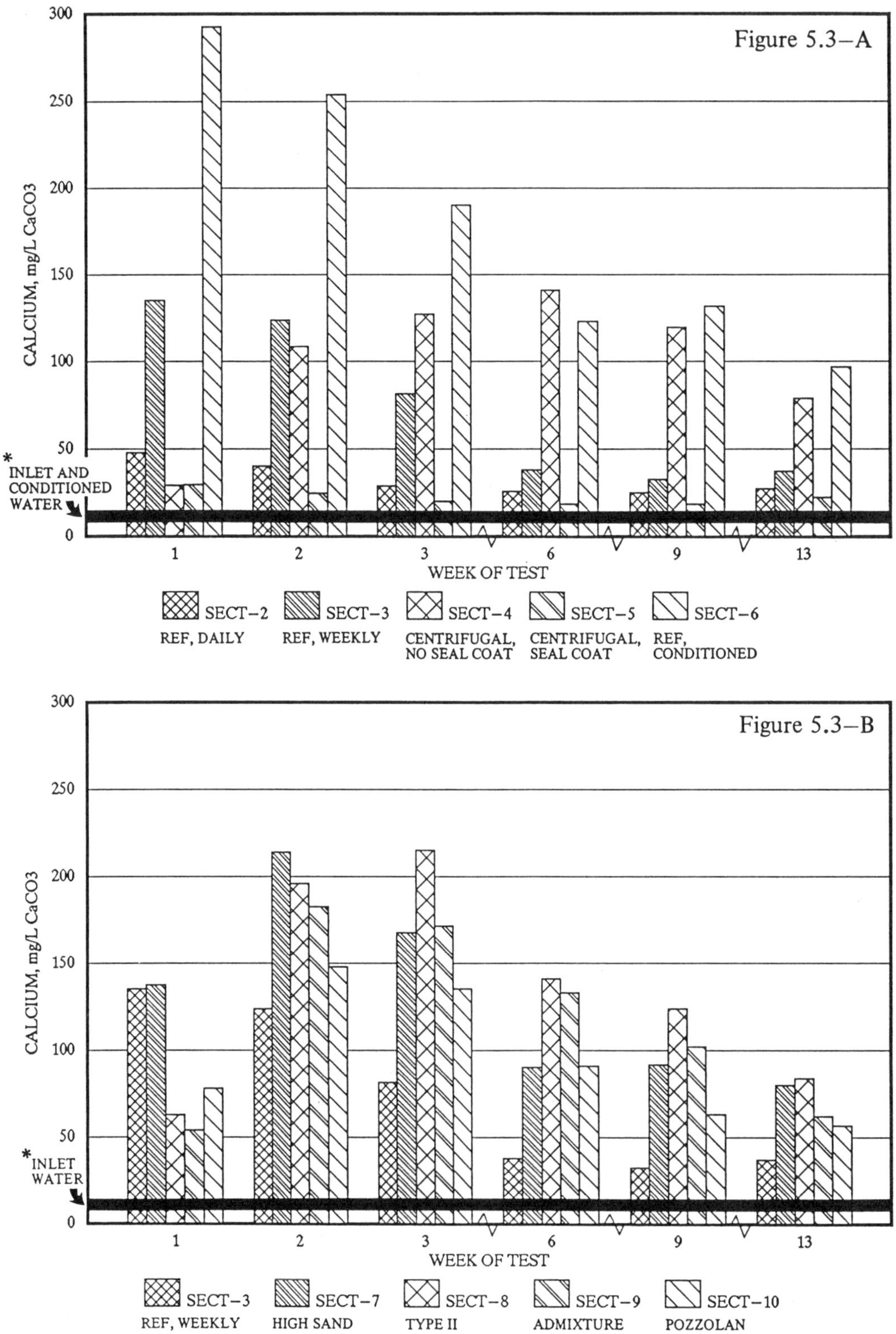

*SECT–1 (Reference Mortar, Continuous Flow) had the same composition as the Inlet Water.

Figure 5.3 South Central Connecticut RWA calcium results

performance of section 4 with the performance of section 8, which was also lined with a Type II cement mortar, but by the (reference) drag trowel method. The "A" series of figures presents section 4 data as the middle bar of each group; directly below, the middle bars of the "B" series of figures present section 8 data.

For the first few weeks, the centrifugally lined section performed much better than the drag trowel section. The pH after the first week was 11.9 in the centrifugally lined section, versus 12.4 in the drag trowel section; the alkalinity after the first week was 280 mg/L $CaCO_3$, versus 680 mg/L $CaCO_3$; and the calcium was 30 mg/L $CaCO_3$, versus 60 mg/L $CaCO_3$. The performances of the two sections began to converge toward the sixth week, and by the ninth week, water quality differences were minimal. While the centrifugally lined section still had superior results, both sections finished the 13-week test with a pH of 11.1 to 11.3, alkalinity of 80 to 110 mg/L $CaCO_3$, and calcium of 80 to 85 mg/L $CaCO_3$.

To summarize, the centrifugally lined section without seal coat performed better than the drag trowel section, particularly in the first few weeks of the field test. Both sections used a Type II cement mortar and were subjected to weekly flushing.

Seal Coat

The effect of an asphaltic seal coat can be determined by comparing sections 4 and 5 in Figures 5.1A, 5.2A, and 5.3A. These two sections, adjacent on the figures, were both centrifugally lined, flushed weekly, and used Type II cement in the linings. The only difference between the two was the seal coat on section 5.

The seal coat had a major impact on water quality. The seal-coated section's pH stabilized near 10.0, more than 1 unit lower than the pH of the centrifugally lined section without a seal coat. Similarly, the alkalinity from the seal-coated section stabilized significantly lower than alkalinity from the section without seal coat, 30 versus 80 mg/L $CaCO_3$. The calcium from the seal-coated section approached the inlet water concentration of 13 mg/L $CaCO_3$; in contrast, the water from the section without a seal coat had high calcium values throughout the test, stabilizing near 100 mg/L $CaCO_3$.

For all three key parameters each week of the test, the seal-coated section outperformed all other weekly flushed sections. Water contacting the seal coat instead of the cement-mortar lining leached very little lime.

It is not clear, however, if the seal-coated section complied with the requirements in ANSI/AWWA Standard C104 that the coated pipe must not impart more than 25 mg/L of either hardness or alkalinity to water held for 24 hours. The test matrix called for flushing the water only once per week. During the first week, calcium increased 24 mg/L as $CaCO_3$, complying with the standard. Alkalinity increased from 12 mg/L $CaCO_3$ in the inlet water to 204 mg/L in the first flush, an average increase of 27 mg/L per day, implying that the section did not comply.

Water-Conditioning Chemicals

South Central Connecticut RWA adds a zinc metaphosphate corrosion inhibitor (1 part Zn to 7.5 parts metaphosphate, as P_2O_5, total dose 2 mg/L) as part of its normal corrosion control program, and all sections except section 6 received zinc metaphosphate-conditioned water. Section 6, in contrast, received water that had been conditioned with the same conditioner used by the Marin MWD (1:3 zinc orthophosphate, zinc dose 0.3 mg/L). Comparing the results of section 3 and section 6 tests allows an evaluation of the relative effect of the two conditioning programs.

Zinc orthophosphate addition lowered the pH of the conditioned water to a steady average of 6.8, compared with 7.2 in the inlet water. The chemical also lowered alkalinity—to 10 versus 14 mg/L $CaCO_3$—but had little effect on calcium, which, like the inlet water, was 13 mg/L $CaCO_3$. The "A" series presents section 6 data as the right-hand bars in each group, with its own baseline of conditioned water.

The pH of the weekly flushed zinc orthophosphate-conditioned water was among the highest of all sections in the field test, 12.6 after the first week, stabilizing at 11.3 to 11.4. Similarly, poor results occurred with alkalinity—920 mg/L $CaCO_3$ after the first week, decreasing to 110 mg/L $CaCO_3$ at the end of the test. Calcium concentrations were also high, 290 mg/L $CaCO_3$ after the first week and 100 to 130 mg/L $CaCO_3$ late in the test.

Zinc concentrations were very low in the first and subsequent flushes from section 6, about 0.03 mg/L Zn, compared with 0.24 mg/L in the conditioned water, which charged the pipe section. This suggests that zinc precipitated or was adsorbed in the section. The orthophosphate concentration, 0.09 mg/L PO_4 in the conditioned water, was reduced to below 0.03 mg/L after 1 week in the conditioned-water section, also suggesting removal by precipitation or adsorption.

The poor performance of section 6—with respect to pH, alkalinity, and calcium—may have been caused more by a lining defect than by the zinc orthophosphate conditioning chemical. As described later, a patch of lining had flaked away in this section, which may have been the source of intense, localized lime leaching. Therefore, the effects of zinc orthophosphate on cement-mortar lining corrosion are unclear, based on this field test.

Mortar Composition

The "B" series of figures demonstrates the effects of various cement-mortar compositions on lining corrosion. All sections in these figures were lined by the drag trowel method and were flushed weekly during the field test. The left-hand bar represents the section lined with the reference mortar; next is the high sand-to-cement ratio mortar; next the mortar with Type II cement; then the mortar with a water-reducing admixture; and on the right-hand side is the mortar with pozzolanic material.

The reference mortar performed the best of these sections, with the lowest pH, alkalinity, and calcium concentrations for all but the first 2 weeks of the test. The Type II cement section performed the worst, finishing the 13-week test with a pH of 11.3 (0.4 units higher than the reference section), alkalinity of 110 mg/L $CaCO_3$ (40 mg/L $CaCO_3$ higher), and calcium of 85 mg/L $CaCO_3$ (50 mg/L $CaCO_3$ higher). Of the intermediate performers, the pozzolan section results compared with the reference section results most closely, though at the start of the test it had the highest pH and alkalinity values. The strong finish of the pozzolan section may be related to the tendency of pozzolanic mortars to continue developing strength longer than mortars without pozzolan. The high sand and admixture sections finished with similar pH and alkalinity values, though the admixture section began the test with high alkalinity releases. Calcium losses were always large from the high sand mortar, nearing the concentrations of the Type II cement section.

Calcium concentrations after the first week were relatively low for most of the sections, rising much higher the second week. It may be that during the first week, calcium concentrations were indeed so high initially that the waters precipitated out calcium carbonate, thus reducing the dissolved calcium residual.

These results show that the reference mortar had the least impact on water quality at South Central Connecticut RWA. The pozzolan section was the next best performer. The Type II cement section had the worst results.

Other Inorganic Water Quality Parameters

While water quality evaluations focused on the lime-related parameters of pH, alkalinity, and calcium, researchers also analyzed silica, zinc, magnesium, phosphate, iron, and manganese. This section summarizes the results of these analyses. Table 5.3 presents the inlet water quality with respect to these parameters.

The silica results presented in Figures 5.4A and 5.4B show that the water in many sections leached silica compounds from cementitious linings. The centrifugally lined section without seal coat had the highest silica concentration after the first week, 95 mg/L SiO_2, and the concentration had not decreased to the inlet silica concentration even by the end of the test. Sections containing Type II cement, admixture, and pozzolan also had high early silica concentrations. Except for the first week, the calcium concentrations of these sections followed similar profiles (see Figure 5.3A), as might be expected because cementitious compounds are primarily silica and calcium. As postulated previously, it may be that the profiles were, in fact, similar during the first week as well, but that initially high calcium concentrations precipitated out of solution.

Zinc and magnesium, only 0.13 and 1.9 mg/L, respectively, in the inlet water, probably precipitated or were adsorbed in many of the sections, based on lower concentrations in the flushed waters. Phosphate precipitates probably also formed, though the orthophosphate concentration was not measured in the inlet water. The zinc orthophosphate additive in the conditioned water produced baseline zinc and orthophosphate concentrations of 0.19 and 0.09 mg/L, respectively. In the conditioned-water section, these concentrations were 0.02 and less than 0.03 mg/L, respectively, indicating that precipitation or adsorption occurred. Manganese concentrations were always below the detection limit. As discussed later, only the linings in the continuous-flow sections had visible scale deposits, so the precipitation described above appeared to have no impact on cement-mortar lining corrosion.

Table 5.3 South Central Connecticut RWA inlet water quality for other inorganic parameters

Parameter	Concentration (mg/L)
Iron, Fe	0.02
Magnesium, Mg	1.9
Manganese, Mn	0.01
Silica, SiO_2	7.6
Total phosphate, PO_4	0.35
Zinc, Zn	0.13

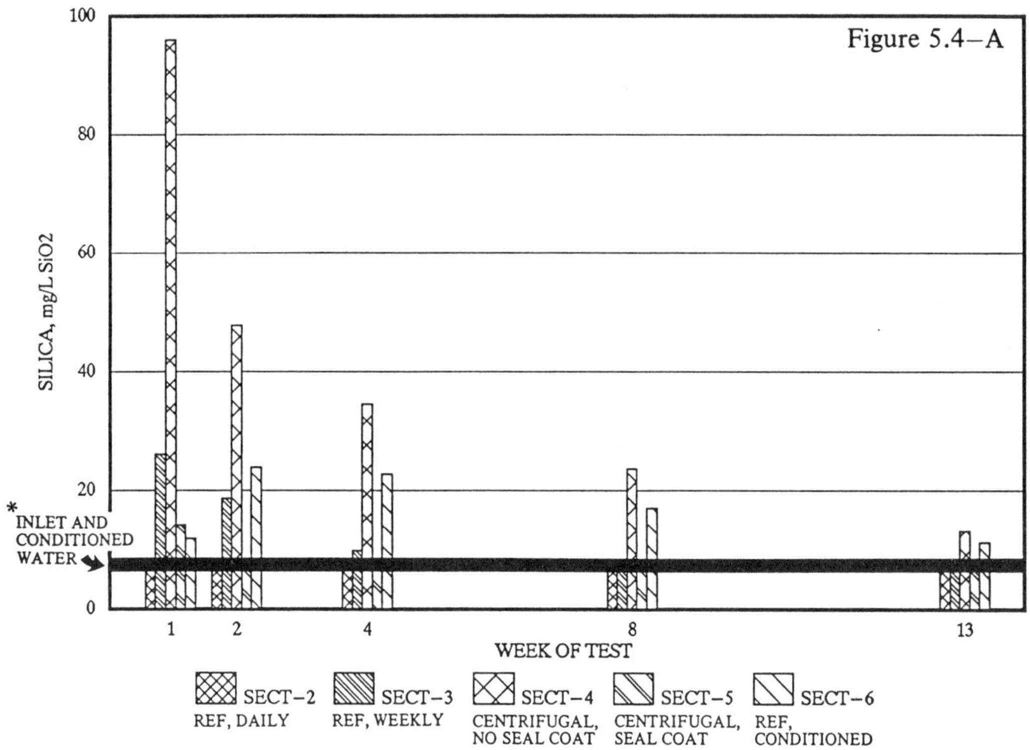

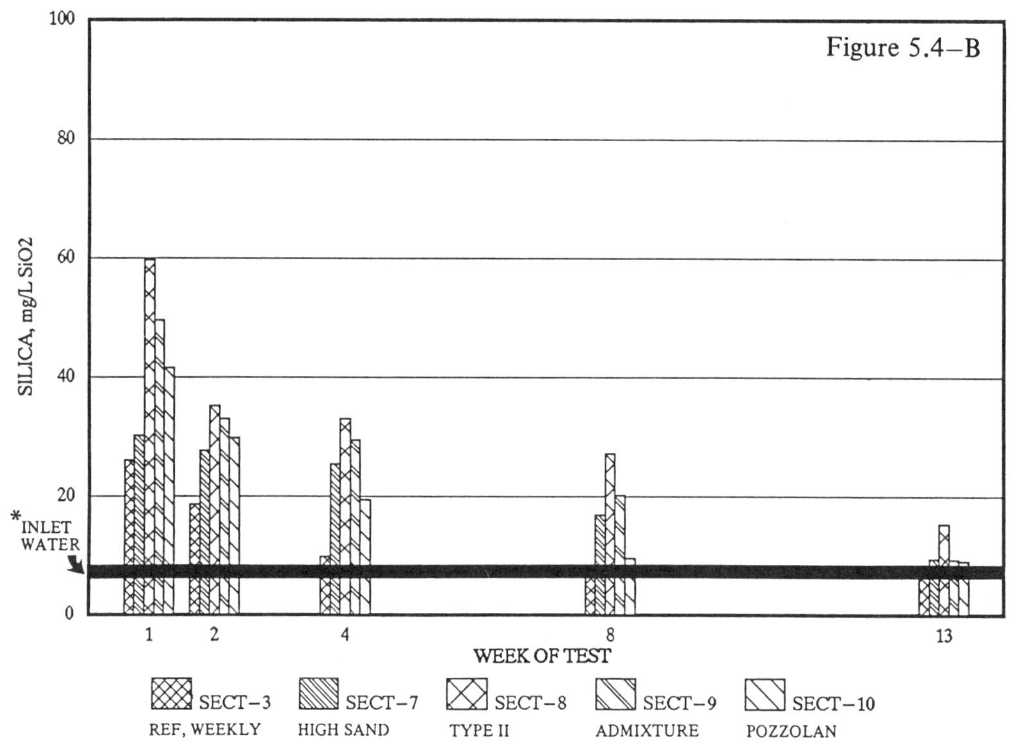

*SECT–1 (Reference Mortar, Continuous Flow) had the same composition as the Inlet Water.

Figure 5.4 South Central Connecticut RWA silica results

Iron was present in the inlet water at 0.02 mg/L and generally remained at that level in most of the waters analyzed. The occasional higher concentrations, up to 0.1 mg/L or more, typically occurred in sections that, in post-test inspections, were found to have rusty spots, usually at the ends of the sections, where metal was inadvertently exposed.

Marin MWD Water Quality Results

Table 5.4 summarizes pH, alkalinity, and calcium results at the Marin MWD field test site. Inlet water quality is described below, with respect to these three key parameters, followed by discussions of the corrosion factors investigated—flow regime, lining method, seal coating, conditioning chemicals, and mortar composition. The results for silica and the other monitored inorganic parameters are presented, after which the organic compound results are discussed.

Inlet Water

For the first 9 weeks of the field test, the Marin MWD inlet water maintained steady pH, alkalinity, and calcium values of 7.8, 71 mg/L $CaCO_3$, and 38 mg/L $CaCO_3$, respectively. This water, with an LI of –0.6, was less aggressive than the South Central Connecticut RWA inlet water. On July 6, 1990, Marin MWD stopped dosing finished (inlet) water with caustic, as part of a THM-control test program. This action lowered the inlet water pH and alkalinity to 6.9 and 55 mg/L $CaCO_3$ for the final 4 weeks of the test, creating a more aggressive water with an LI of –1.7. Table 5.5 presents average inlet water values for pH, alkalinity, and calcium, along with standard deviations for the periods before and after July 6, 1990.

Marin MWD used two sources of water during the test period—Nicassia Lake and Kent Lake. These two waters were very similar with respect to pH, alkalinity, and calcium. During the first 9 weeks, the source water changed from Nicassia Lake to Kent Lake (on May 18, 1990), then back to Nicassia Lake (on June 18, 1990), with little impact on these three parameters. As described later, the main difference between the two source water qualities was silica concentration.

Flow Regime

Marin MWD operated the same three flow regimes as South Central Connecticut RWA—continuous flow (section 1), water held stagnant with daily flush (section 2), and water held stagnant with weekly flush (section 3). Each of these sections was lined with the reference cement mortar by the drag trowel method.

Water flowed through the continuous-flow section at 9 gpm (0.57 lps), an 11-second hydraulic detention time. This brief contact with the lining caused no change in the pH, alkalinity, and calcium values, as shown in Table 5.5. However, these parameters did increase in the sections with daily and weekly flushes.

pH. Figure 5.5A demonstrates the effect of flow regime on pH at Marin MWD. Continuous-flow pH values did not increase during the brief lining contact. Daily flushed pH values stabilized at 9.8 to 10.0, about 2 units higher than the inlet values. Weekly flushed values were about 3 units higher than inlet values, stabilizing at 10.8 to 11.0. These stabilized pH values were not greatly affected by the decreased inlet pH during the final 4 weeks of the test.

Table 5.4 Summary of Marin MWD field test results

Section	pH	Alkalinity (mg/L $CaCO_3$)	Calcium (mg/L $CaCO_3$)
Inlet water (fed to all sections but section 6)	Steady at 7.8 for first 9 weeks; from middle of 10th week until the end of test pH averaged 6.8*	Steady at 72 for first 9 weeks, from middle of 10th week until the end of test alkalinity averaged 55*	Steady at 35 to 40 throughout test, except for low values (28 and 16) on 6th and 10th weeks, respectively
Alternatively conditioned (fed to section 6 only)	Stable at 8.0 to 8.3	Stable at 55 to 60	Stable at 40
1. Reference, continuous flow	Steady at 7.8 for first 9 weeks, from middle of 10th week until the end of test pH averaged 6.8	Steady at 72 for first 9 weeks, from middle of 10th week until the end of test alkalinity averaged 56	Steady at 35 to 40 throughout test (low value of 15 on 10th week)
2. Reference, daily flush	11.2 after first day, down to 10.0 by end of 2nd week, stable at 9.8 to 10.0 until 10th week after which pH averaged 9.3 to 9.7	160 after first day, decreased slowly to a stable 90 to 110, low low value of 85 on 13th (final) week	60 to 70 throughout test
3. Reference weekly flush	11.5 after first week, stable at 10.8 to 11.0 until 10th week after which pH averaged 10.6	190 after first week, stabilized at 130 after 9 weeks	70 after first week, remained at 45 to 55 through 7th week, rose to near 70 after 10th week
4. Centrifugally lined, no seal coat, weekly flush	11.6 after first week, decreased rapidly to 10.0 to 10.2, stabilized at 9.8 during final weeks	250 after first week, decreased rapidly and stabilized at 100 to 110 after 7 weeks	60 to 80 through 10th week after which calcium increased to near 90
5. Centrifugally lined, seal coat, weekly flush	10.9 after first week, stabilized at 8.7 after 11 weeks	160 after first week, stabilized at 70 (inlet water value) after 11 weeks	Stable from start at 40 (inlet water value)
6. Reference, alternative conditioning chemicals, weekly flush	11.6 after first week, decreased slowly to 10.5, may not have stabilized during test	270 after first week, decreased to 115, still decreasing at end of test	100 after first week, stabilized at 60 to 65 after 7 weeks
7. High sand-to-cement ratio, weekly flush	12.0 after first week, stable at 10.8 to 11.0 until 10th week after which pH averaged 10.6	620 after first week, stabilized at 120 after 6 weeks	380 after first week, decreased to 50 to 60 through 10th week after which calcium increased to 80
8. Type II cement, weekly flush	11.8 after first week, stable at 10.8 until 10th week after which pH averaged 10.6 to 10.7	280 after first week, stabilized at 140 after 7 weeks	70 after first week, decreased to 55, then increased slowly to 95, may not have stabilized during test

*On July 6, 1990, Marin MWD stopped adding caustic (NaOH) to the water as part of a test program for THM control.

Table continues next page

Table 5.4 Summary of Marin MWD field test results (continued)

Section	pH	Alkalinity ($mg/L\ CaCO_3$)	Calcium ($mg/L\ CaCO_3$)
9. Water-reducing admixture, weekly flush	11.8 after first week, stable at 10.7 to 10.9 until 10th week after which pH averaged 10.6	280 after first week, stabilized at 130 to 140 after 8 weeks	100 after first week, remained at 60 to 70 through 10th week, after which calcium increased to 85
10. Pozzolan, weekly flush	12.0 after first week, stable at 10.8 to 11.0 until 10th week, after which pH averaged 10.6	570 after first week, stabilized at 130 after 9 weeks	300 after first week, decreased to 50 then increased slowly to 85, may not have stabilized

Table 5.5 Marin MWD inlet versus continuous-flow water quality

Water description	pH		Alkalinity ($mg/L\ CaCO_3$)		Calcium ($mg/L\ CaCO_3$)	
	Average	Standard deviation	Average	Standard deviation	Average	Standard deviation
May 1 through July 6, 1990						
Inlet water	7.82	0.16	71	4.1	38	3.6
Continuous flow (section 1 water)	7.82	0.16	71	4.6	39	5.8
July 6 through July 30, 1990						
Inlet water	6.85	0.09	55	1.2	33	10.1
Continuous flow (section 1 water)	6.87	0.10	57	0.9	34	10.7

Alkalinity. The effect of flow regime on alkalinity is shown in Figure 5.6A. The inlet water alkalinity was 72 mg/L $CaCO_3$ during most of the test, as was the continuous-flow alkalinity. Again, only the daily and weekly flushed sections exhibited water quality effects. The daily flushed alkalinity was 160 mg/L $CaCO_3$ after the first day, decreasing to a stable 90 to 110 mg/L $CaCO_3$ by the end of the test. The weekly flushed alkalinity was 190 mg/L $CaCO_3$ after the first week, stabilizing at 130 mg/L $CaCO_3$ after 9 weeks.

Calcium. Figure 5.7A shows how flow regime affected calcium concentrations. The continuous-flow calcium concentrations matched inlet levels, steady at 35 to 40 mg/L $CaCO_3$. Daily flushed calcium concentrations also were steady, though higher at 60 to 70 mg/L $CaCO_3$. In the weekly flushed section, the calcium concentration was 70 mg/L $CaCO_3$ after the first week, then it decreased to 45 to 55 mg/L $CaCO_3$, where it remained through the tenth week. During the final weeks

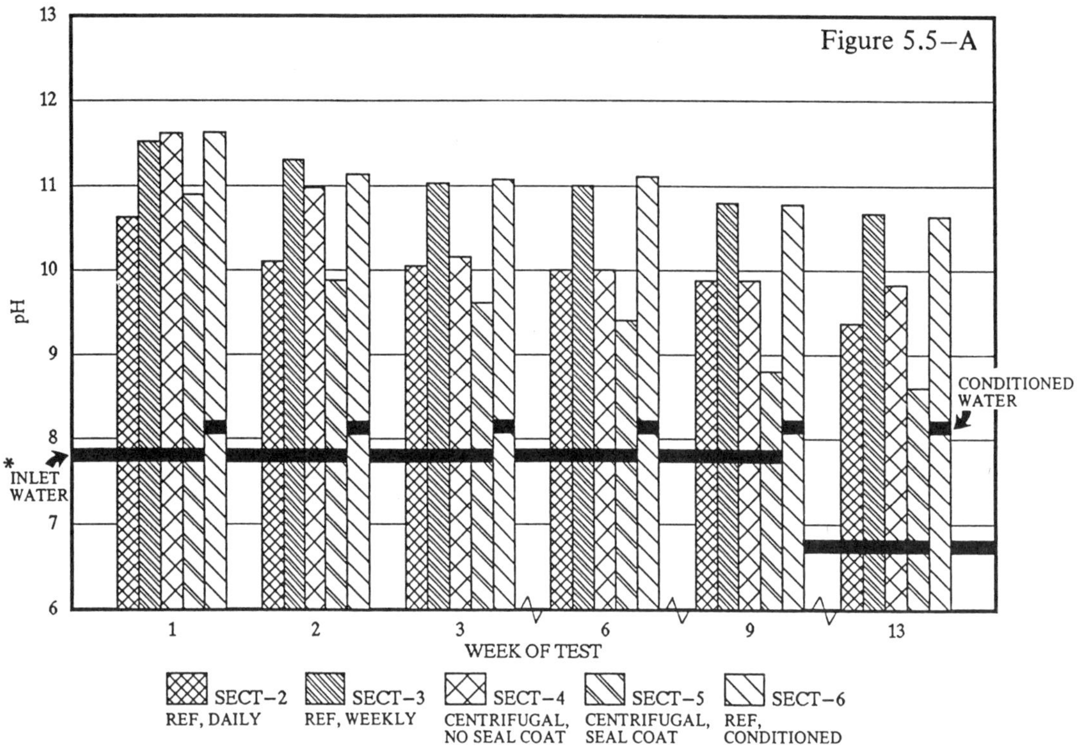

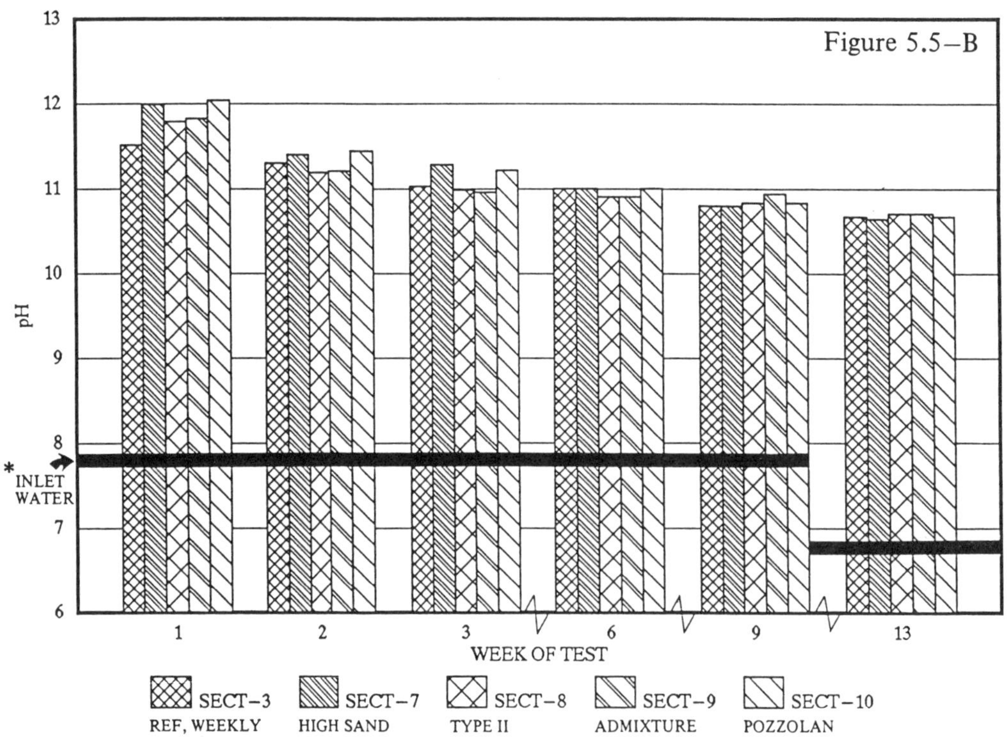

*SECT–1 (Reference Mortar, Continuous Flow) had the same composition as the Inlet Water.

Figure 5.5 Marin MWD pH results

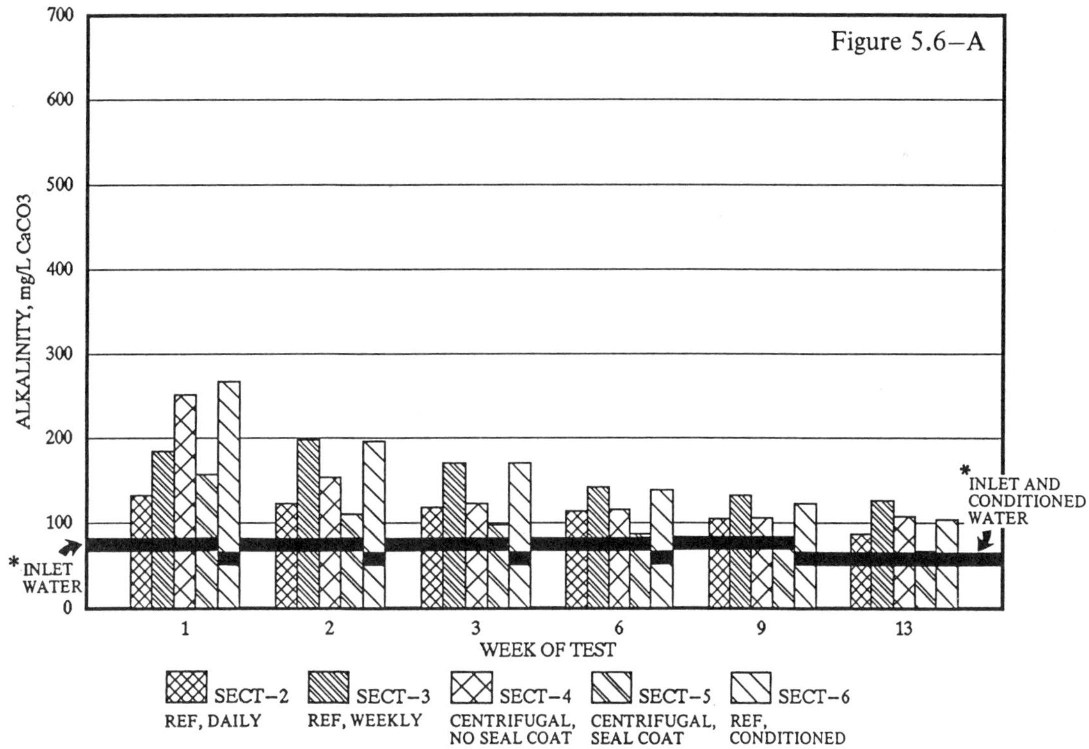

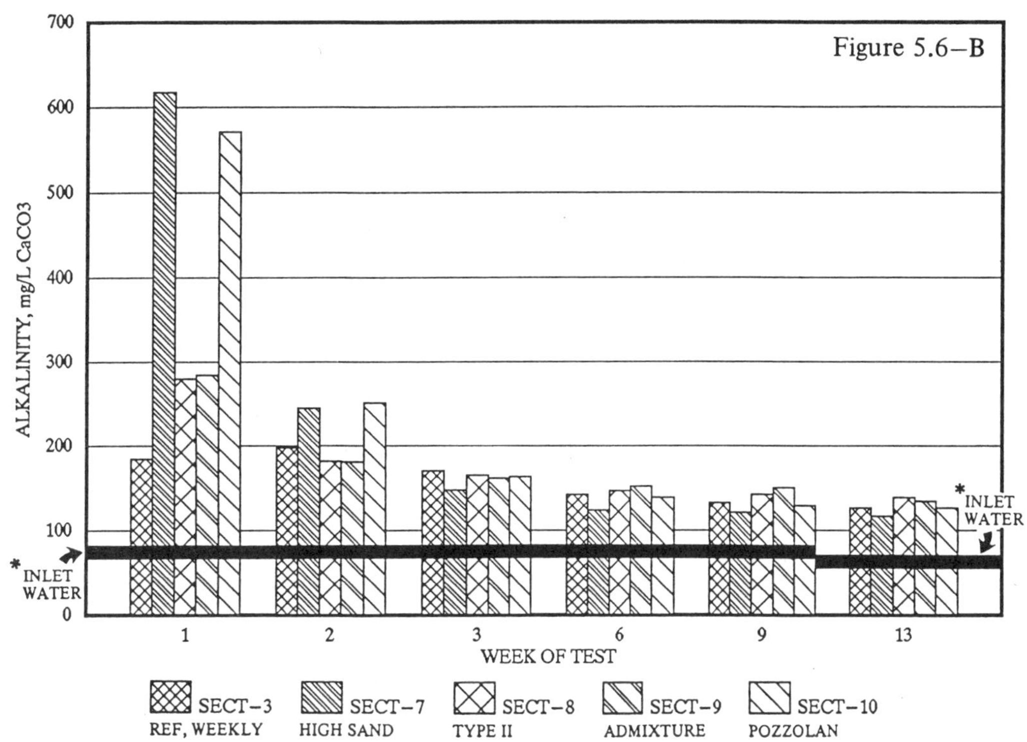

*SECT–1 (Reference Mortar, Continuous Flow) had the same composition as the Inlet Water.

Figure 5.6 Marin MWD alkalinity results

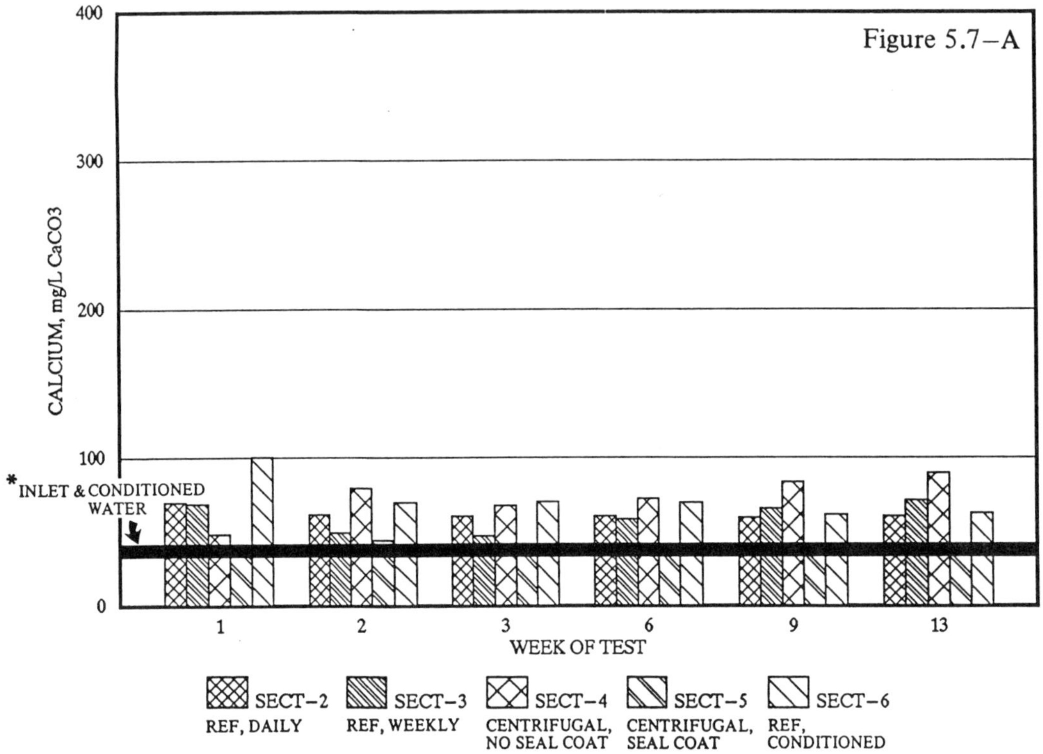

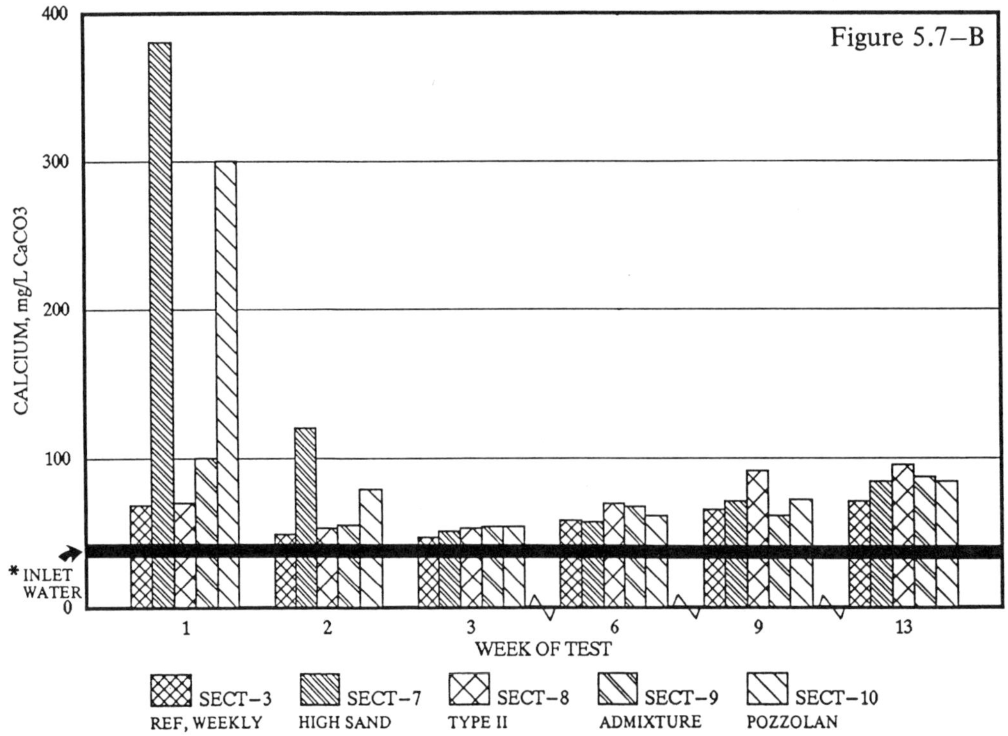

*SECT–1 (Reference Mortar, Continuous Flow) had the same composition as the Inlet Water.

Figure 5.7 Marin MWD calcium results

of the test, the weekly flushed calcium concentrations rose to 70 mg/L $CaCO_3$. These higher concentrations may have occurred because the inlet water was more aggressive, with a lower pH, during this period.

Summary. Marin MWD flow regime results support the results from South Central Connecticut RWA—longer contact with the lining causes greater water quality impacts from leached lime. In addition, the change in Marin MWD inlet water quality during the tenth week produced interesting results. When the water became more aggressive (lower LI), it leached more lime from the linings. This was particularly evident from the calcium concentrations, which exhibited steady increases in most sections throughout the final 4 weeks of the test.

Drag Trowel Versus Centrifugal Lining

The effect of lining method can be determined by comparing the results from sections 4 and 8. Section 4 was lined by the centrifugal method with Type II cement (see the middle bars of the "A" series of figures). Section 8 was lined by the drag trowel method, also with Type II cement (see the middle bars of the "B" series).

For the first 2 weeks, the two sections had similar pH values, though the centrifugally lined section was 0.2 units lower. As the test progressed, however, the gap widened, with the centrifugally lined pH values stabilizing near 10.0, while the drag trowel pH values remained higher at 10.7. The centrifugally lined alkalinities stabilized at 100 to 110 mg/L $CaCO_3$, 30 to 40 mg/L lower than the 140 mg/L $CaCO_3$ in the drag trowel method. Calcium concentrations were essentially the same, finishing the test near 90 mg/L $CaCO_3$.

To summarize, centrifugal lining had fewer water quality impacts than drag trowel lining, particularly with respect to pH. The centrifugally lined section without seal coat had the best results of all Marin MWD pipe sections, except the seal-coated, centrifugally lined section.

Seal Coat

Examination of section 5 results in Figures 5.5A, 5.6A, and 5.7A shows that the seal-coated section outperformed all other sections throughout the test for all three key parameters. The pH was 10.9 after the first week, stabilizing at 8.7, 1 unit lower than the next best section. The alkalinity was 160 mg/L $CaCO_3$ after the first week, stabilizing at 70 mg/L $CaCO_3$, approaching the inlet water alkalinity. The calcium concentration was stable throughout the test at 40 mg/L $CaCO_3$, essentially the same as the inlet water calcium. These results support the findings of the South Central Connecticut RWA field test—a seal coat minimizes lime leaching into the water.

With respect to complying with the alkalinity and hardness release limitations of ANSI/AWWA Standard C104, the seal-coated section of Marin MWD had results superior to those of South Central Connection RWA's section 5. During the first week, calcium increases were near zero, thus complying with the standard. Alkalinity increases were from 72 mg/L $CaCO_3$ in the inlet water to 157 mg/L in the first flush. This represents an average daily increase of only 12 mg/L, versus the 25-mg/L limit for any 24-hour period. However, without knowledge of the day-to-day increases, it is not certain whether the section complied with the standard.

Water-Conditioning Chemicals

Marin MWD adds zinc orthophosphate conditioner (1 part zinc to 3 parts orthophosphate, zinc dose 0.3 mg/L) as part of its normal conditioning program. All sections except section 6 received zinc orthophosphate-conditioned water. Section 6, in contrast, received a water dosed with a combination polyphosphate–orthophosphate conditioner (0.8 mg/L total PO_4). The polyphosphate in this chemical sequesters dissolved iron from corroding pipe, which prevents the iron from forming the precipitate that leads to red-water customer complaints. Comparing the results of section 3 and section 6 tests allows evaluation of the relative effects of the two conditioning programs.

The pH of the feedwater to section 6 averaged 7.9, the alkalinity averaged 60 mg/L $CaCO_3$, and the calcium averaged 40 mg/L $CaCO_3$. It had an LI of –0.6, similar to that of the inlet water (during the first 9 weeks). At the field test dose of 0.8 mg/L total PO_4, the total phosphate concentration was not too different from that of the inlet water (1.0 mg/L PO_4). However, the section 6 feedwater contained primarily polyphosphate, while the inlet water contained only orthophosphate.

Water from section 6 generally had pH, alkalinity, and calcium concentrations slightly lower than the water from section 3. The pH results were a close match, each near 11.5 after the first week and finishing the test at 10.6. The water from section 6 had poor alkalinity and calcium results after the first week, then finished better than the water from the reference section. The alkalinity for the water from section 6 was 115 mg/L $CaCO_3$ at the final week and still was decreasing, compared with the stabilized value of 130 mg/L $CaCO_3$ for water from the reference section. During the final 4 weeks, the water from section 6 had calcium values of 60 to 65 mg/L $CaCO_3$ and seemed to be less affected by the newly aggressive inlet water than the section 3 water, which finished the test with a calcium concentration rising from 65 to 70 mg/L $CaCO_3$.

In summary, polyphosphate–orthophosphate conditioning chemicals, used to control red water, may have a slightly beneficial impact on cement-mortar lining corrosion in low-flow systems.

Mortar Composition

The "B" series of figures demonstrates the effects of various cement-mortar compositions on lining corrosion. There was very little difference in results for these five linings.

The reference section had a pH of 11.5 after the first week, lower than the other sections (11.8 to 12.0), but all sections stabilized at 10.6 to 10.7. During the first week, the high sand lining and the pozzolan lining lost large amounts of alkalinity and calcium. However, results grouped together by the third week. The Type II cement section finished the test with calcium concentrations about 10 to 15 mg/L $CaCO_3$ higher than the other sections.

All the various mortar compositions performed about the same at Marin MWD. Changes involving sand proportion, admixture, and pozzolan had little effect on water quality. The Type II cement section stabilized at slightly higher pH, alkalinity, and calcium values than sections with other mortar compositions.

Other Inorganic Water Quality Parameters

In addition to pH, alkalinity, and calcium, researchers also analyzed silica, zinc, magnesium, phosphate, iron, and manganese. Table 5.6 presents the inlet water quality with respect to these parameters. The silica concentration varied from 3 to 9 mg/L SiO_2 in the inlet water, because water sources varied during the test. The different water sources did not affect the other parameters significantly.

Figures 5.8A and 5.8B present the silica results. They are complicated by the fact that changing source waters caused the inlet silica concentrations to vary throughout the test, as indicated in the figures. During the first 2 weeks, silica concentrations exceeded the inlet concentration of 3 mg/L SiO_2 to a large extent in the centrifugally lined section without seal coat, and to a lesser extent in the sections with high sand, admixture, and pozzolan. The source water was different in the third week, with a higher inlet silica concentration. All sections produced waters with silica concentrations lower than in the inlet water, indicating silica removal by precipitation or adsorption. In the eighth and thirteenth weeks of the test (the only other weeks these additional inorganic parameters were measured), the Marin MWD source water was again the original water, and silica concentrations in all sections essentially matched inlet concentrations.

Zinc, magnesium, and orthophosphate probably precipitated or were adsorbed in all weekly flushed sections, based on the low concentrations in the water samples compared with the inlet water concentrations. Magnesium precipitation probably occurred in the early weeks when pH values commonly exceeded 11, but not thereafter.

Iron and manganese concentrations were below the detection limit in the inlet water and nearly all section samples. Iron concentrations near 0.4 mg/L occurred in the centrifugally lined sections, which later inspection revealed to have rusty patches around the pipe flanges.

Organic Water Quality Parameters

Marin MWD investigated the potential of the seal coat to release organic compounds. Having observed that the seal-coated section performed well with respect to lime leaching, the utility is considering changing its existing pipe lining standard, which currently specifies no seal coat. With this in mind, Marin MWD also investigated reasons to possibly not use seal coat. Marin MWD chose this opportunity to measure possible leaching of PAH, THMs, and volatile organic compounds.

Table 5.6 Marin MWD inlet water quality for other inorganic parameters

Parameter	Concentration (mg/L)
Iron, Fe	<0.05
Magnesium, Mg	9.4
Manganese, Mn	<0.03
Orthophosphate, PO_4	0.96
Silica, SiO_2	5.63
Zinc, Zn	0.38

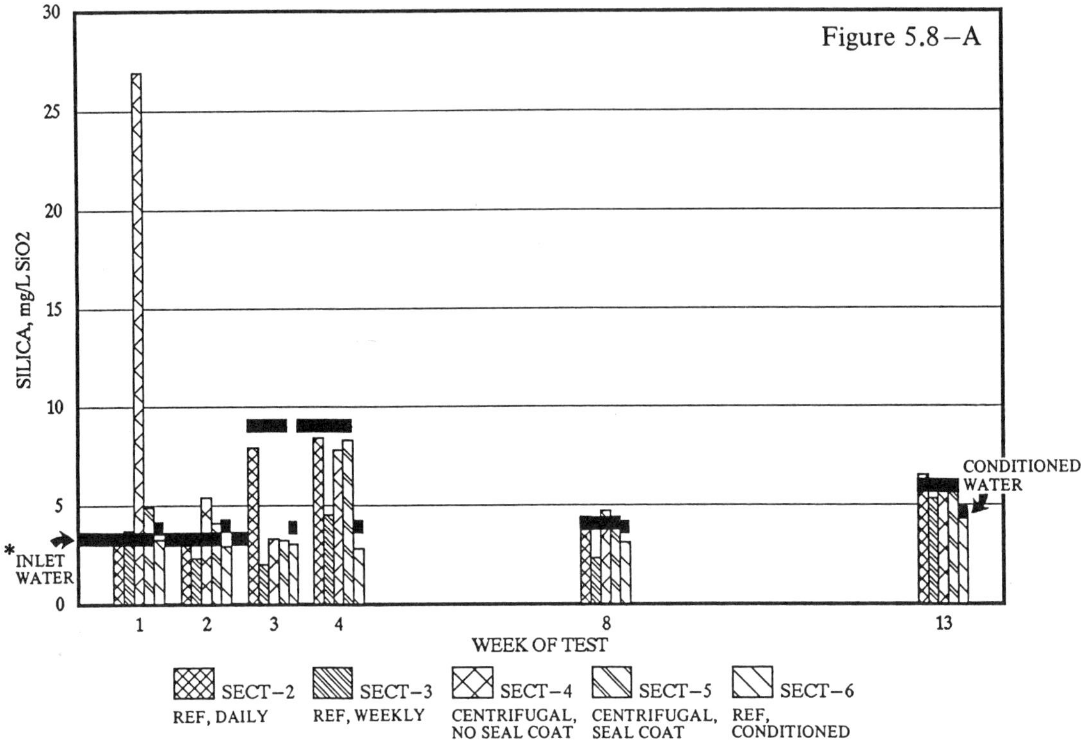

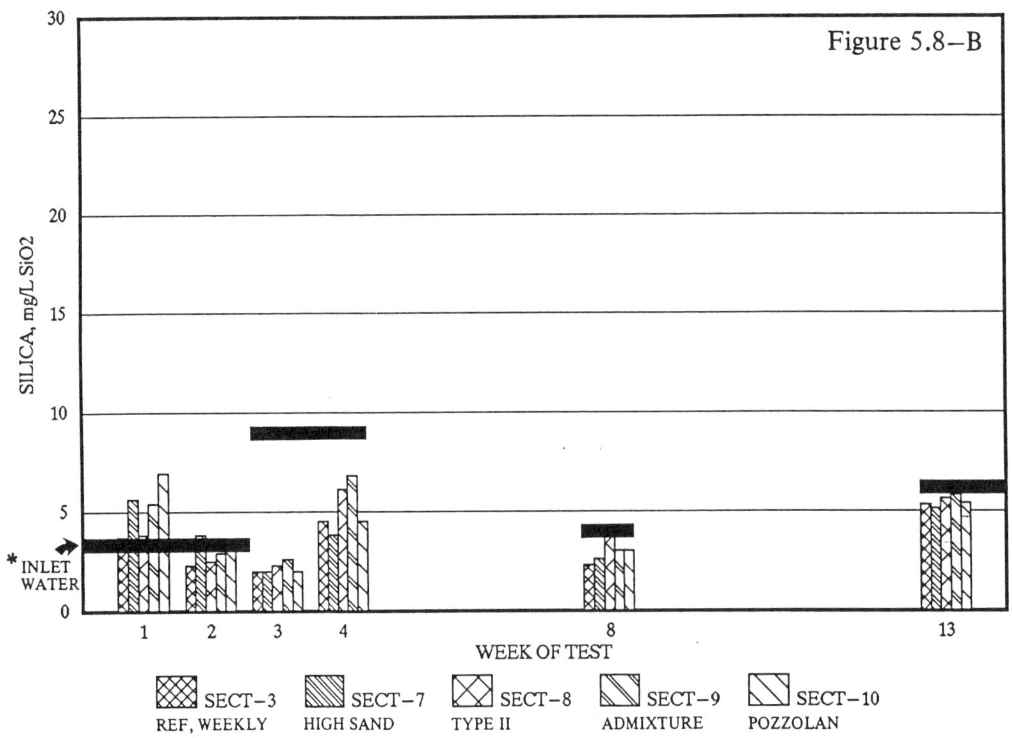

*SECT–1 (Reference Mortar, Continuous Flow) had the same composition as the Inlet Water.

Figure 5.8 Marin MWD silica results

On the last day of the test, July 30, 1990, Marin MWD sampled waters from the continuous-flow section (section 1) and the two centrifugally lined sections—without seal coat (section 4) and seal coated (section 5). The continuous-flow water was intended to represent inlet water quality. The utility analyzed for THMs, PAHs, and volatile halogenated and aromatic organic compounds. All samples were analyzed within 1 day of sampling. In addition, the continuous-flow water was analyzed for THMs again after 4 days, approximately the detention time in the water distribution system. By performing this delayed analysis, Marin MWD could estimate the potential of the water to form THMs just prior to reaching customers. Table 5.7 summarizes the results of the organic analyses. Results for THMs, PAHs, and volatile organic compounds are discussed below.

THMs. Total THMs in the continuous-flow water were 29 µg/L (micrograms per liter) when first analyzed and 58 µg/L after 4 days. Both concentrations were less than the MCL (maximum contaminant level) of 100 µg/L recommended by EPA (US Environmental Protection Agency).* Chloroform was the principal THM. During the additional time between analyses, THM concentrations in the continuous-flow sample approximately doubled, both collectively and individually. Based on these results, it is reasonable to assume that THM concentrations in the other two sections, each measured about 65 µg/L within 1 day of sampling, would be higher if they also had been measured after 4 days. They might double, as did the continuous-flow concentrations, or they might increase by a similar increment (for example, total THM in section 4 might increase by 29 µg/L). If THM concentrations were to double after 4 days, they would exceed the MCL, demonstrating an organic water quality problem associated with cement-mortar lining corrosion.

Trihalomethane formation is generally believed to be accelerated by high pH values. The higher pH values in the daily and weekly flushed sections probably caused the higher THM concentrations.

Apparently, a pH increase from 6.8 to 8.7 is significant with respect to THM formation. However, total and individual THM concentrations were similar in the two centrifugally lined, weekly flushed sections, 65 to 66 µg/L, even though the seal-coated section stabilized at a lower pH than the section without seal coat, 8.7 versus 9.8.

PAHs. Where they exist, MCLs for PAHs are very low, less than 1 µg/L, because of health concerns. As shown in Table 5.7, only one PAH, benzo(a)pyrene, currently has an MCL, though EPA is considering MCLs for six others.

Analyses revealed very low PAH concentrations. Only four PAHs were detected, none of which is currently regulated or expected to be regulated soon. The highest PAH concentrations occurred in water from the seal-coated section. Phenanthrene was measured at 0.16 µg/L; anthracene, fluorene, and pyrene were measured at concentrations less than 0.1 µg/L. Thus, this water did not violate proposed EPA PAH standards.

Volatile Organic Compounds. Tetrachloroethylene (PCE) was the only volatile halogenated or aromatic organic compound detected in any of the waters. It was not present in the continuous-flow water. However, it was measured in water

*The 1974 SDWA (Safe Drinking Water Act) and its 1986 revisions require EPA to develop MCLs for various drinking water contaminants.

Table 5.7 Marin MWD organic compound concentrations (μg/L)

Parameter	Maximum contaminant level (MCL)	Section 1 reference mortar, continuous flow: Analyzed within 1 day of sampling	Section 1 reference mortar, continuous flow: Analyzed 4 days after sampling	Section 4 centrifugally lined, no seal coat: Analyzed within 1 day of sampling	Section 5 centrifugally lined, seal coat: Analyzed within 1 day of sampling
THM					
Chloroform	NA*	19	35	41	39
Bromoform	NA	<0.5	<0.5	<0.5	<0.5
Bromodichloromethane	NA	7	18	21	17
Dibromochloromethane	NA	3	5	4	9
Total THM	100	29	58	66	65
PAH					
Anthracene	NA	<0.0001	—†	0.0004	0.006
Fluorene	NA	<0.0005	—	<0.0005	0.076
Phenanthrene	NA	0.002	—	0.004	0.16
Pyrene	NA	<0.005	—	0.006	0.022
Benzo(a)pyrene	0.2	<0.001	—	<0.001	<0.001
Benzo(a)anthracene	0.1‡	<0.001	—	<0.001	<0.001
Benzo(b)fluoranthene	0.2‡	<0.002	—	<0.002	<0.002
Benzo(k)fluoranthene	0.2‡	<0.001	—	<0.001	<0.001
Chrysene	0.2‡	<0.001	—	<0.001	<0.001
Dibenzo(a,h)anthracene	0.3‡	<0.002	—	<0.002	<0.002
Indeno(1,2,3-c,d)pyrene	0.4‡	<0.005	—	<0.005	<0.005
Acenaphthene	NA	<0.003	—	<0.003	<0.003
Acenaphthylene	NA	<0.02	—	<0.02	<0.02
Benzo(g,h,i)perylene	NA	<0.02	—	<0.02	<0.02
Fluoranthene	NA	<0.01	—	<0.01	<0.01
Napthalene	NA	<0.04	—	<0.04	<0.04
Volatile organic compounds					
Tetrachloroethylene	5	<0.5	<0.5	17	80
Others§	§	<0.5	<0.5	<0.5	<0.5
Stabilized field test pH		6.8	6.8	9.8	8.7

*NA means no MCL for this parameter.

†Not analyzed.

‡The EPA is considering this MCL for this parameter.

§Marin MWD analyzed for all parameters identifiable by EPA Method 502—volatile organic compounds in water. These analytical procedures identify approximately 50 volatile halogenated and aromatic compounds, none of which has an MCL greater than 0.5 μg/L, the detection limit. No compound other than tetrachloroethylene (PCE) was found.

from the centrifugally lined section without seal coat at 17 μg/L and in water from the centrifugally lined, seal-coated section at 80 μg/L. These concentrations are both in excess of the MCL of 5 μg/L.

The high result from the centrifugally lined section without seal coat (17 μg/L) was unexpected, because the major anticipated PCE source (the seal coat) was not present. The result suggested that PCE might have also been leached from other construction materials (for example, PVC flanges, rubber gaskets, or epoxy used to repair initial mortar damage). These materials were present in both the seal-coated section and the section without seal coat. Assuming PCE was leached from these "other" materials to a similar extent in the seal-coated section, they contributed only 20 percent (17 out of 80 μg/L) of the PCE in the water. This suggests seal coat was the major PCE contributor in the seal-coated section.

To pinpoint the source(s) of PCE in the section without seal coat, Marin MWD personnel exposed PVC nipples, a gasket, and epoxy patches to the same test conditions that had produced the high PCE concentrations in the previous tests. They substituted PVC nipples for PVC flanges because they could find no closed containers large enough to hold the flanges. The test conditions were:

- One week exposure to Marin MWD water in liquid-filled closed containers.
- Each material in its own separate container.
- Water pH adjusted to 9.8.
- Same ratio of material surface area to water volume as in the test sections.

After exposure, they analyzed the liquid fractions and a blank sample for PCE. The results were as follows:

- Blank, <0.5 μg/L PCE.
- Water contacting PVC nipples, <0.5 μg/L PCE.
- Water contacting epoxy patch, 6.5 μg/L PCE.
- Water contacting gasket, 6.9 μg/L PCE.

The sum of the contributions (14 μg/L) is approximately the same as the PCE concentration (17 μg/L) previously found in the section without seal coat, confirming that results are reasonably reproducible. The epoxy patch and gasket contributed equally to the PCE concentration in the section without seal coat.

As indicated previously, the high PCE contribution of the seal coat was unexpected. According to the Ductile Iron Pipe Research Association (DIPRA 1991), all domestic ductile iron pipe manufacturers have had their seal-coated pipe tested by a standard protocol (National Sanitation Foundation Standard [NSF] 61 for Drinking Water System Components, Health Effects 1988), and these pipes have been in compliance with requirements for potable water contact. DIPRA also indicated the major supplier of asphaltic coating has had its material tested, and this material was in compliance.

The DIPRA information raises the question about whether the extraction protocols of this study differed sufficiently from the extraction protocols of NSF Standard 61 to produce an artificially high PCE result. This question is difficult to answer, since NSF holds confidential which of its test methods are used for certifying specific products and the reasoning that leads to certification, or withholding of certification. Inspection of NSF Standard 61 protocols most likely to be used (section 3.0, Appendix B [Pipes and Related Products] and section 4.0, Appendix B [Protective (Barrier) Materials]) does not clearly indicate whether either would be more likely or less likely to leach PCE than the protocol used in this study. All that can be concluded is that the PCE results of this study differ from the results reported by DIPRA.

Lining Effects

At the end of the field tests, the pipe sections were inspected to see if the test conditions affected the physical structure of the linings. A set of control sections that had not been subjected to test conditions were inspected. The control sections provided baseline information about the original lining condition.

Physical inspections checked linings for color, hardness, sand loss, scales, and defects. Any scales were analyzed for silica, calcium, zinc, magnesium, phosphate, iron, manganese, and total inorganic carbon (the sum of H_2CO_3, HCO_3^-, and CO_3^{-2}). Selected sections were also analyzed for calcium loss with lining depth.

Physical Inspection

The Marin MWD seal-coated lining (section 5) was black and had a dusting of rust-colored powder. It had no obvious scales or defects, but did give off a mildly unpleasant odor. The South Central Connecticut RWA seal-coated lining (section 5) had developed isolated rust-colored spots over a small percentage of its surface. Some spots had built up into tubercules. This lining had no obvious odor. Hardness and sand loss were not relevant to seal-coated linings.

Most pipe sections without a seal coat had similar physical characteristics. In general, these linings were:

- White to gray in color.
- Hard (could not be scratched with a thumbnail).
- Not subject to sand loss when rubbed with a wire brush.
- Without visible scale.
- Without serious fabrication defects.

There were exceptions to the above description of the linings without a seal coat. The exceptions were two scale formations, one lining with soft areas, and two defects in lining fabrication. These conditions are described below.

Scale Formation

Scale formed on the linings of both continuous-flow sections (section 1). This result is in keeping with the common observation at water utilities that deposition of protective scales is more likely to occur in high flow regimes. For example, it is difficult to precipitate a layer of calcium carbonate in dead ends for the purpose of corrosion control.

At South Central Connecticut RWA, a rust–brown scale covered the lining in section 1. Table 5.8 presents the analysis of the scale material, based on sampling approximately 2 mils* (0.05 mm) scraped from the innermost (water-side) portion of the lining. The scale was composed primarily of silica, calcium, and carbonate, with smaller amounts of phosphate, magnesium, and iron.

---

*1 mil = 0.001 inches.

A gray–brown scale on the Marin MWD continuous-flow section (section 1) did not cover the lining completely. As shown in Table 5.8, the scale contained silica, calcium, and carbonate, with lesser amounts of phosphate, magnesium, and iron. It had a composition similar to the scale in the South Central Connecticut RWA continuous-flow section, though with a higher silica concentration.

While only the linings in the two continuous-flow sections had visible scale, water quality results indicate that zinc, magnesium, and phosphate had precipitated or were adsorbed in many sections during the field tests. Therefore, this sampling and analytical technique was used to analyze the innermost portions of linings in three Marin MWD sections (sections 3, 6, and 7). Waters in these sections had exhibited significant reductions in these parameters. As shown in Table 5.8, the composition of the innermost portions (first 2 mils) of these linings is about the same as the compositions of the scales, consisting primarily of silica, calcium, and carbonate, but showing evidence of zinc, magnesium, and phosphate precipitation or adsorption.

Soft Areas

Only one lining had soft areas—the Marin MWD conditioned-water section (section 6). The lining in this section had a few small slimy patches, which could be scraped off with a thumbnail. Analysis revealed their compositions to be like the scales described above, consisting primarily of silica, calcium, and carbonate (see Table 5.8).

Table 5.8 Composition of scales or other portions of linings for test sections (percent)

	South Central Connecticut RWA		Marin MWD			
Parameter	Section 1: Reference mortar, continuous flow*	Section 6: Reference mortar, conditioning chemicals, weekly flush†	Section 1: Reference mortar, continuous flow*	Section 3: Reference mortar, weekly flush‡	Section 6: Reference mortar, conditioning chemicals, weekly flush‡	Section 7: High sand mortar, weekly flush‡
Silica, SiO_2	25	42	51	34	44	33
Calcium, Ca	5	5	7	7	12	6
Carbonate, CO_3	10	5	12	12	8	12
Zinc, Zn	<0.01	<0.02	0.3	<0.01	<0.01	<0.01
Magnesium, Mg	1	2	0.3	.7	2	0.5
Phosphate, PO_4	2	3	1	1	0.2	0.7
Iron, Fe	0.4	1	1	0.3	0.5	0.3
Manganese, Mn	0.01	0.05	0.1	0.02	0.02	0.01
Unidentified substances	43.4	41.9	27.3	45.0	33.3	47.5
Total	100.0	100.0	100.0	100.0	100.0	100.0

*Scale observed on this section.

†No scale observed. Results pertain to loose material thought to have dislodged from defective portion of lining.

‡No scale observed. Results pertain to innermost 2 mils of lining (approximately).

Defects

Two sections had physical lining defects that appeared to originate from the fabrication process. These defects may have affected test results.

A patch of lining had flaked away in the South Central Connecticut RWA conditioned-water section (section 6). The patch, several square inches in area and about 1⁄16 in. deep, did not penetrate to metal. However, it was prone to sand loss, and there was loose material gathered in the bottom of the section. Table 5.8 presents the analysis of the loose material—mostly silica, calcium, and carbonate, with a relatively high percentage of phosphate (3 percent). This damaged area may have contributed to the large water quality effects noted previously—significant increases in pH, alkalinity, and calcium concentrations relative to inlet water values.

The Marin MWD centrifugally lined section without a seal coat (section 4) had a half-dollar-sized holiday in the lining, exposing bare metal. The portion of the lining below the holiday (about 1 ft [0.3 m] of pipe) had a rusty discoloration. This defect may have affected water quality results, though this section performed better than all but the seal-coated section at the Marin MWD field test. Exposure to metal presumably caused the high iron concentrations measured early in the test in the water from this section.

Calcium Loss

Loss of calcium from the linings is evidence of cement corrosion and deterioration. The major calcium loss is presumably caused by lime leaching; leaching of other compounds (calcium silicates, for example) might also occur, probably to a more limited degree. The materials testing laboratory analyzed the calcium content of thin layers of mortar cut from selected pipe sections, using the procedures described in "Scale Composition, Calcium Loss, and pH Effects," chapter 4. Laboratory personnel made four cuts per section, each cut right over the other. The first cut was 5 mils (0.13 mm) deep, the second 10 mils (0.25 mm) deep, the third 15 mils (0.38 mm) deep, and the fourth 20 mils (0.51 mm) deep. The combined cut depths summed to 50 mils (1.27 mm), approximately 30 percent of the total lining thickness.

Figure 5.9 plots the measured calcium content of each cut (expressed as CaO) against the depth to the midpoint of each cut for the reference mortar exposed to three different flow regimes at South Central Connecticut RWA. The greater the exposure to water, the lower the calcium content. Figure 5.10 shows similar results for Marin MWD. These figures show that increased flows accelerate calcium leaching, even though they do minimize changes to water quality by diluting the leached components.

Although one would expect control sections (sections not exposed to flowing or flushed water) to have higher calcium contents than the exposed section, this is apparently not always so. For example, the sections flushed weekly appear to have higher calcium contents than the control sections (see Figures 5.9 and 5.10). No explanation is apparent for this unexpected behavior. Also unexpected was that the measured calcium values in the weekly flushed sections would be higher than the calcium values estimated from a material balance on the fresh mortar components (32 percent, as CaO).

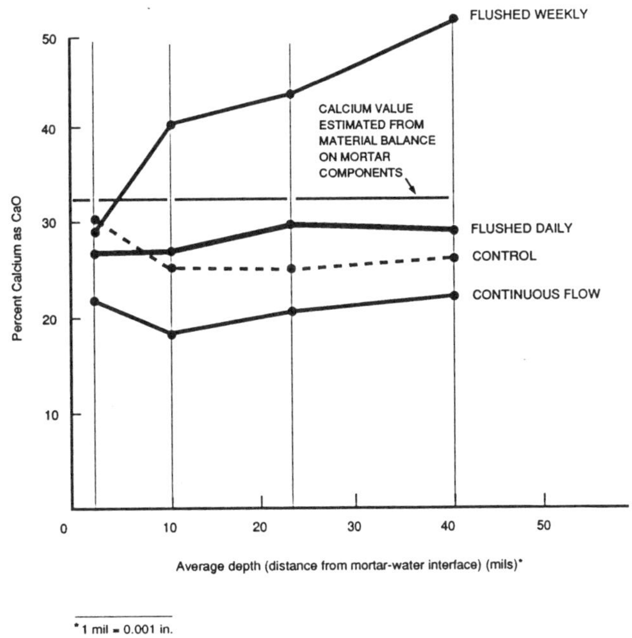

Figure 5.9 Calcium content (as CaO) versus average mortar depth for reference mortar for three different flow regimes—South Central Connecticut RWA

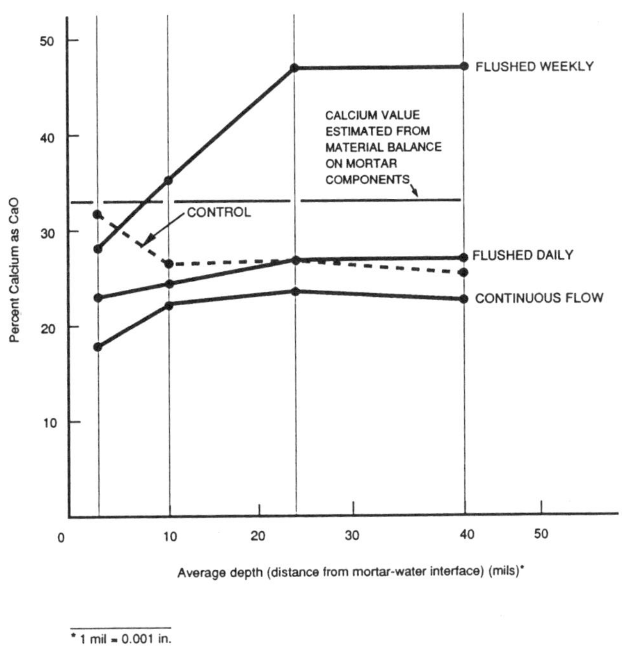

Figure 5.10 Calcium content (as CaO) versus average mortar depth for reference mortar for three different flow regimes—Marin MWD

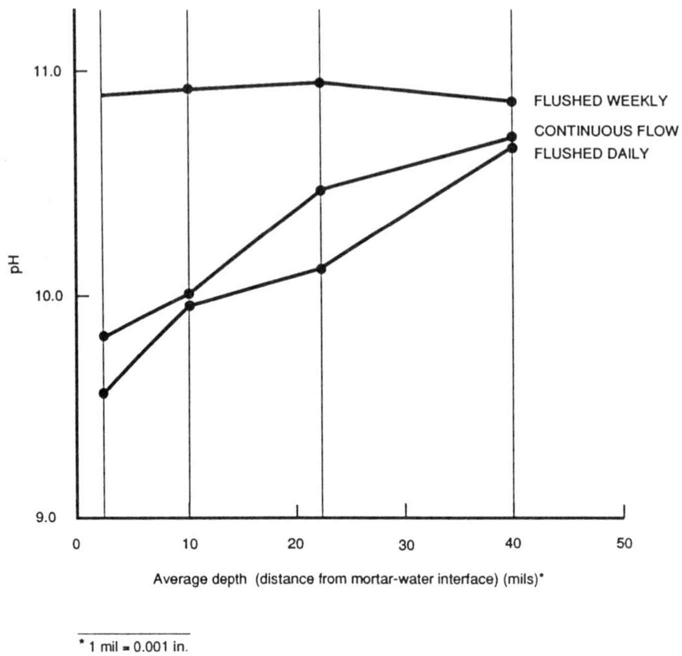

Figure 5.11 pH of slurries made by mixing cuttings from different depths with deionized water—South Central Connecticut RWA

pH Effect

Laboratory personnel took portions of the cuttings collected for the South Central Connecticut RWA calcium analyses (described in "Calcium Loss," above) and slurried them with deionized water. They mixed 0.18 g (grams) of solid sample with 0.100 L of water for 10 minutes, settled the slurry for 5 minutes, then measured the pH of the supernatant. Figure 5.11 shows supernatant pH as a function of mortar depth. Lower pH values suggest samples relatively depleted of basic (high pH) materials, presumably lime.

The section with minimum exposure to water (weekly flushed) produced the slurry with the highest pH as would be expected. The pH values of slurries from the daily flushed and continuous-flow sections were lower. The weekly flushed section had a relatively flat pH profile, in contrast to its initially steep calcium profile. The other two sections had steeper pH profiles than their corresponding calcium profiles.

Examination of Aged Pipe Sections

Following completion of the formal test program (the results of which are described on the previous pages), South Central Connecticut RWA personnel dug up three sections of cement-mortar-lined pipe, which had been in service for different periods of time. They then shipped 1-ft (3-m) long pieces of each section to the materials testing laboratory for physical inspection, scale analyses, and measurements of calcium loss and pH effects. The objective of these tests was to determine how the physical and chemical characteristics of linings exposed to low or intermittent flows changed upon exposure to water for extended, but different, periods of time. This kind of information could not be obtained from the formal test program, because the period of exposure in the formal test program was relatively short (3 months).

The distinguishing characteristics of the aged pipe sections were:

- Section A—A 6-in. (15.2-cm) pipe that had been in service 1 year.
- Section B—A 6-in. (15.2-cm) pipe that had been in service 6 years.
- Section C—An 8-in. (20.3-cm) pipe that had been in service 12 years.

All three pipes had been lined in situ and all had been installed in areas of the distribution system with low or intermittent flows.

Physical Inspection

Grayish, mineral-like deposits covered the water-side surfaces of sections B and C. These sections had been in service the longest (6 and 12 years). The section B deposit was about 50 mils (1.3 mm) thick, while the section C deposit was about 75 mils (1.9 mm) thick. Both deposits were relatively uniform in thickness. They were not as hard as the cement mortar. The deposits not only were different in appearance than the underlying mortar, they appeared to be physically distinct. In places near the end of the sections, the deposits fell away from the mortar in strips if slightly disturbed. The deposit in section C was smoother than the deposit in section B. Section A, in contrast to sections B and C, had no visible deposits.

The mortar linings (deposits excluded) in all three sections were not uniformly thick, varying in thickness from ⅛ to ¼ in. (3.0 to 6.3 mm) around the perimeter of the pipe.

Scale Analyses

Using the procedures described in "Scale Composition, Calcium Loss, and pH Effects," chapter 4, the analysts cut to mid-depth of the deposits in sections B and C, then analyzed the cuttings. Because section A had no visible scale, the analysts cut away only the first 5 mils (0.1 mm) of mortar in section A, then analyzed those cuttings. Table 5.9 shows analyses of the cuttings for sections A, B, and C. As with the sections from the test rig (see Table 5.8), silica, calcium, and carbonate were prime components of sections A, B, and C cuttings. However, the iron contents of sections A, B, and C cuttings were much greater than the iron contents of the cuttings from the test rig, and the zinc contents were also higher.

Calcium Loss

Following the scale analyses, analysts cut away and discarded the remainder of the deposits from sections B and C. They then cut thin layers of the underlying

Table 5.9 Scale analyses for sections A, B, and C (percent)

Parameter	Section A (1 year)	Section B (6 years)	Section C (12 years)
Silica, SiO_2	41	19	29
Calcium, Ca	9.8	9.4	6.1
Carbonate, CO_3	8.7	9.6	12.1
Zinc, Zn	0.8	1.4	1.9
Magnesium, Mg	0.3	0.5	0.7
Phosphate, PO_4	0.6	1.4	2.0
Iron, Fe	9.8	12.3	13.6
Manganese, Mn	0.08	0.03	0.38
Unidentifiable substances	28.9	46.4	34.2
Total	100.0	100.0	100.0

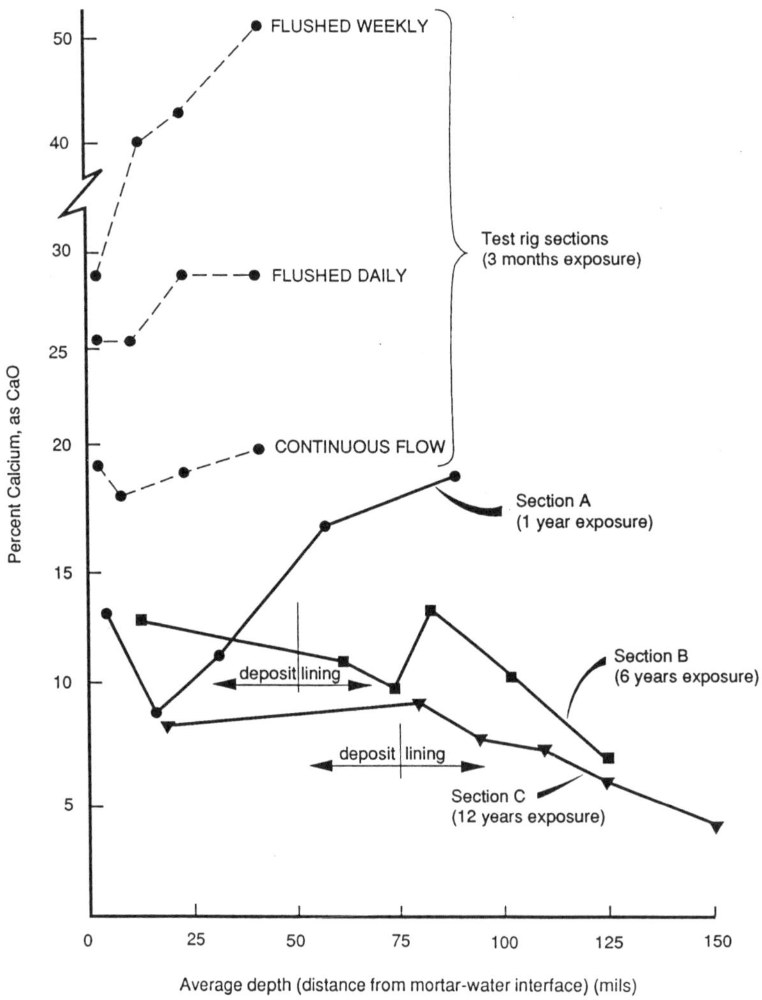

Figure 5.12 Calcium content (as CaO) versus average mortar depth for various exposure times—South Central Connecticut RWA

mortar from all three sections (five cuts per section) and analyzed the cuttings for calcium content. The combined cuts summed to approximately 80 percent of the thickness of the mortar layer. Figure 5.12 (heavy lines) shows the calcium profiles for these sections. The far-left data point of the profiles for sections B and C shows the calcium content of the deposits. The vertical line shows the location of the interface between the deposits and the mortar.

The calcium content of the least-exposed mortar (1 year) was, for the most part, higher than the calcium content of a sample exposed for 6 years. However, the calcium content of the 6-year-old sample was only marginally higher than that of the 12-year-old sample. The calcium profiles of the older sections are flatter than the calcium profile of section A.

For comparison, calcium profiles for some of the linings from the South Central Connecticut RWA test rig sections are shown on the same figure (dashed lines). They have much higher calcium contents than the aged pipe.

The data in Figure 5.12 indicate that calcium leaches rapidly in the early periods of exposure, with the leaching slowing down as time progresses. This conclusion is supported by the water quality data. The result is not surprising. The leaching rate naturally slows down as surface calcium becomes depleted. The deposits formed after extensive exposure also serve to protect the mortar against further leaching.

pH Effect

Laboratory personnel took portions of the cuttings collected for the calcium loss measurements and slurried them with deionized water, then settled the samples, and measured the pH of the supernatants. They used the same proportions of mortar to water (0.18 g to 0.100 L), and the same contacting procedures and contact times as in the pH effect experiments with mortars from the test rig. Figure 5.13 plots supernatant pH against mortar depth (heavy lines). The mortar from the section exposed for 12 years produced about the same result overall as mortar from the section exposed for 6 years. The mortar exposed for 1 year produced a higher pH solution. For comparison, pH results for some of the mortars from the South Central Connecticut RWA test rig sections (exposure times 3 months) are superimposed on the figure (dashed lines). Figure 5.13 confirms the results of Figure 5.12; namely that lime leaching slows down as exposure time increases.

Summary

The field tests at both sites produced waters that contained dissolved lime, as indicated by elevated pH, alkalinity, and calcium values relative to inlet water values. When water flowed continuously past the lining, water quality impacts were negligible. Daily and weekly flushed waters showed significant pH, alkalinity, and calcium increases. Weekly flushed pH values were similar to those at the surveyed utilities that reported cement-mortar lining corrosion problems.

The degree of corrosion in the field tests was related to the aggressiveness of the inlet water. Increases in pH, alkalinity, and calcium at the South Central Connecticut RWA field test site, where the inlet water had a very low LI (–2.5),

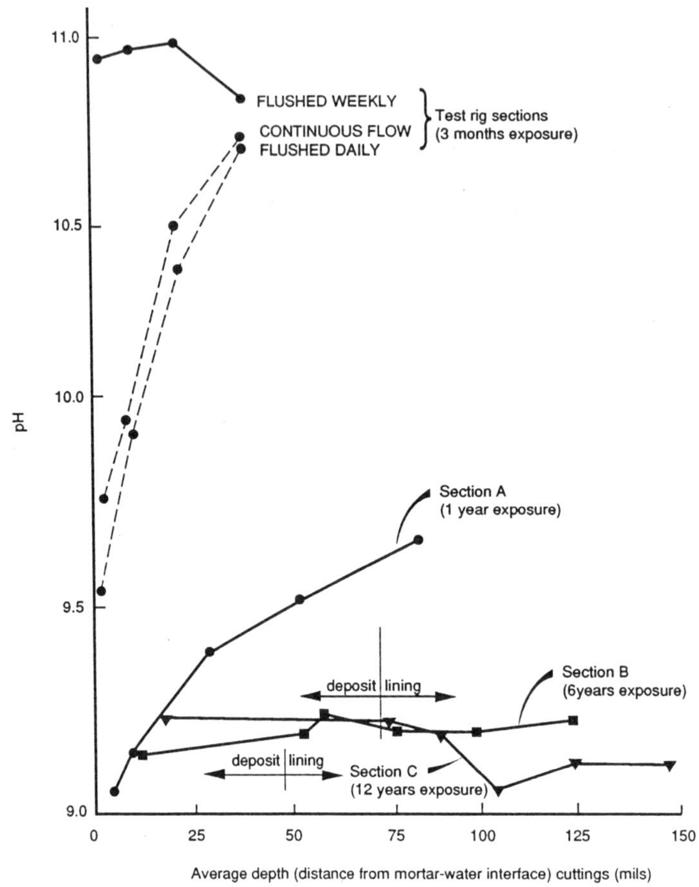

Figure 5.13 pH of slurries made by mixing cuttings from different depths with deionized water—South Central Connecticut RWA

were much larger than increases at the Marin MWD site, particularly during the first week of the test. The South Central Connecticut RWA site also leached more silica. Moreover, when Marin MWD's inlet water became more aggressive during the final 4 weeks (because the utility stopped adding caustic), the calcium concentration rose in the flushed waters from most sections (the seal-coated section maintained its calcium value). Although the LI correlated with these trends, other water quality parameters, for example, alkalinity, may also be appropriate indicators of the potential to leach cement-mortar compounds.

Steady flushing of the continuous-flow sections diluted the leached lime compounds and kept them from impairing water quality. The waters from the continuous-flow sections had essentially the same composition as the inlet water. However, leaching did occur in these sections. Examinations of calcium concentration versus depth of lining showed that the depth and extent of calcium loss increased with the intensity of flow regime and/or greater exposure to water. Continuous-flow sections exhibited greater calcium loss than daily flushed sections, which in turn lost more calcium than weekly flushed sections. In no case did the leaching appear to weaken the lining.

Of all weekly flushed pipe sections, the centrifugally lined, seal-coated sections clearly performed the best at both sites, with respect to inorganic water quality. However, an analysis for organics at the end of the test by Marin MWD found this water to contain an unexpectedly high PCE concentration. The seal coat appeared to be the major PCE source. This unexpected result may have been an isolated incident since PCE leaching has apparently not been discovered in tests by others. Studies to resolve this question would appear to be worth doing. The seal coat did not leach PAHs.

Tests at Marin MWD measured elevated THM concentrations in waters with high pH caused by contact with cement-mortar linings. An increase in pH from 6.8 to 8.7 substantially increased THM concentrations from 29 to 65 μg/L. However, the increased THM concentrations were still below the MCL for THMs.

At South Central Connecticut RWA, variations in cement-mortar composition affected field test results. The reference mortar performed the best of all sections without seal coat; all other variations of mortar composition produced poorer flushed water quality. Toward the end of the test, the pozzolanic mortar section had results that approached but did not surpass those of the reference mortar section. The admixture, high sand, and Type II cement-mortar sections were consistently worse than the reference mortar section.

The centrifugally lined section without seal coat performed the best of sections without protective coatings at Marin MWD. Among in situ-lined sections, variations in cement mortar made less difference. All sections performed about the same, with the Type II cement-mortar section producing slightly worse results.

The poorest performer at both sites was the in situ-lined Type II cement-mortar section, although its performance was only marginally poorer at Marin MWD. Flushed waters from these Type II sections contained higher concentrations of dissolved lime compounds, particularly calcium, than did waters from the other weekly flushed sections.

There was no apparent correlation between pipe section performance and the characteristics of the hardened mortar used to line a section (see Table 4.2). Neither strength, air content, nor water absorption was a good indicator of trends in cement-mortar lining corrosion.

Lime leaching was slightly lower at Marin MWD when a combination orthophosphate–polyphosphate conditioner was used instead of the utility's usual conditioner. At South Central Connecticut RWA, pH, alkalinity, and calcium concentrations were significantly higher in water from the section that used the zinc orthophosphate conditioner than in water from the section that used South Central Connecticut RWA's usual conditioner (zinc metaphosphate). It is not clear that the differences in performance at South Central Connecticut RWA were caused by the conditioners used. The lining of the section using zinc orthophosphate may have been damaged during fabrication, thus less capable of resisting leaching.

Experiments with pipes removed after 1 to 12 years in low-flow distribution systems indicate that lime continues to leach from the mortar upon prolonged exposure, but the rate of leaching decreases as time goes on. Deposits formed at the mortar/water interface form barriers that may be important in reducing the leaching rate.

Chapter 6

Conclusions and Recommendations

Test results from this study, as well as work done by others, support the following conclusions:

- Water can leach cementitious materials from cement-mortar lining. Extended contact under low-flow conditions can result in the water developing significant increases in concentrations of the leached components, including pH, alkalinity, and calcium. Human responses to these increases vary. Some customers have complained about skin irritations that developed after contact with water from low-flow, cement-mortar-lined systems. Utility personnel are concerned about such complaints, deterioration of the linings, and customers receiving water that differs significantly in quality from water that leaves the treatment plant.
- Extended contact under high-flow conditions appears to accelerate leaching of cement-mortar lining, at least in the short term. Water quality affects are barely noticeable, however, because high flow dilutes the leached materials. It is not clear whether accelerated leaching would continue long term, since protective scales are more likely to form in high-flow areas than in low-flow sections of the distribution system.
- Waters low in pH, alkalinity, and calcium appear to be most aggressive to cement-mortar linings. Some researchers have used alkalinity as an inverse measure of water aggressiveness. The LI might also be a useful approximate indicator, with water aggressiveness increasing as the LI decreases (for LI less than zero).
- Corrosion rates are influenced by cement-mortar composition, the method of lining, and by protective coatings. In this study, centrifugally lined sections protected by coatings of asphaltic seal coat performed best in terms of minimum effect on pH, alkalinity, and calcium. An in situ-lined pipe with a lining composed of a mixture of one part Type I portland cement to one part sand (called the reference mortar) performed the best of all sections without protective coatings at the South Central Connecticut RWA test site. A centrifugally lined section with a lining composed of a mixture of approximately one part Type II portland cement to one part sand performed the best of all sections without protective coatings at the Marin MWD test site. These results with nonseal-coated sections show that performance at one site does not necessarily translate to other sites.

- Differences in performance between different test sections were accentuated at South Central Connecticut RWA, which had the more aggressive water of the two test sites.
- A protective seal coat did not appear to be the source of priority organic pollutants, with one possible exception, tetrachloroethylene (PCE).
- In this short test (3 months duration), water quality effects were more obvious than lining deterioration, in part because water quality effects were measured more often and can be measured more accurately than can parameters associated with lining breakdown. Physical inspection revealed few instances of obvious lining deterioration. The problems that were discovered were probably associated more with fabrication defects than with lining corrosion. Limited chemical analyses indicated depletion of cementitious materials at the mortar/water interface.
- The alternative chemical conditioners used at the two test sites did not appear to reduce cement-mortar leaching in any significant way.

The following recommendations are offered:

- Users of cement-mortar-lined pipe should anticipate deterioration of water quality in low-flow sections of the distribution system. Utilities should regularly flush low-flow sections to prevent buildup of components causing adverse water quality effects.
- In this study's tests, an asphaltic seal coat provided significant protection against cement-mortar leaching. However, analysis of a single water sample from the seal-coated section at the Marin MWD test site revealed PCE in excess of the MCL. This unexpected result may have been an isolated occurrence, since PCE leaching has apparently not been discovered in tests conducted by others. Further research should be done to determine if asphaltic seal coat is a potential source of PCE and other priority organic materials. If so, manufacturers may be able to reformulate the seal coat to eliminate the organic offenders.
- There are currently no known coatings that could be applied in situ to protect cement-mortar lining. The concept is appealing, since there are hundreds of miles of already-installed pipe that could benefit from protection similar to the protection seal coat affords new pipe. Methods for developing and applying such coatings should be considered. To protect water meters and prevent service line plugging, coating devices must be designed to prevent coating materials from entering the service lines.

References

American Society for Testing and Materials. 1978. Standard Test Method for Air Content of Freshly Mixed Concrete by Volumetric Method. ASTM Standard C173-78. Philadelphia: ASTM.

American Society for Testing and Materials. 1984. Standard Test Method for Flexural Strength of Concrete (Using Simple Beam with Third-Point Loading). ASTM Standard C78-84. Philadelphia: ASTM.

American Society for Testing and Materials. 1986. Standard Test Method for Compressive Strength of Cylindrical Concrete Specimens. ASTM Standard C39-86. Philadelphia: ASTM.

American Society for Testing and Materials. 1989. Standard Specification for Portland Cement. ASTM Standard C150-89. Philadelphia: ASTM.

American Society for Testing and Materials. 1990a. Standard Test Method for Slump of Portland Cement Concrete. ASTM Standard C143-90. Philadelphia: ASTM.

American Society for Testing and Materials. 1990b. Standard Specification for Chemical Admixtures for Concrete. ASTM Standard C494-90. Philadelphia: ASTM.

American Society for Testing and Materials. 1990c. Standard Test Methods for Specific Gravity, Absorption, and Voids in Hardened Concrete. ASTM Standard C642-90. Philadelphia: ASTM.

American Society for Testing and Materials. 1991. Standard Specification for Fly Ash and Raw or Calcined Natural Pozzolan for Use as a Mineral Admixture in Portland Cement Concrete. ASTM Standard C618-91. Philadelphia: ASTM.

American Water Works Association. 1989a. AWWA Standard for Cement-Mortar Lining of Water Pipelines—4-in. (100-mm) and Larger—In Place. ANSI/AWWA C602-89. Denver: AWWA.

American Water Works Association. 1989b. AWWA Standard for Cement-Mortar Protective Lining and Coating for Steel Water Pipe—4-in. and Larger—Shop Applied. ANSI/AWWA C205-89. Denver: AWWA.

American Water Works Association. 1990. American National Standard for Cement-Mortar Lining for Ductile-Iron Pipe and Fittings for Water. ANSI/AWWA C104/A21.4-90. Denver: AWWA.

Ashworth, V. 1984. Corrosion: Industrial Problems, Treatment and Control Techniques. Proc. First Arabian Conference on Corrosion, Kuwait.

Babbitt, H.E., and J.J. Doland. 1949. *Water Supply Engineering*, 4th Ed. New York: McGraw-Hill Book Co., Inc.

Benedict, R.L. 1985. Corrosion Protective Properties of Concrete Cylinder Pipe. Paper presented at the Seminar on Corrosion Control, AWWA Distribution System Symposium, Seattle, Wash.

Bonds, R.W. 1989. Cement-Mortar Linings for Ductile Iron Pipe. *Ductile Iron Pipe News*, spring/summer.

Carson, H.Y. 1927. Cement-Lined Water Mains. *Indus. & Engrg. Chemistry*, 19:7.

Council of European Communities. 1980. Council directive of 15 July, 1980, relating to quality of water intended for human consumption (80/778/EEC). Official Journal of the European Communities, Number L229/11, 1980, August.

Dahl, W.R. 1982. Energy Conservation by Cement Mortar Lining. Paper presented at AWWA Annual Conference, Miami Beach, Fla.

Ductile Iron Pipe Research Association. 1991. Personal communications with Richard Bonds, Research/Technical Director. Birmingham, AL: DIPRA.

DVGW W342, 1978. "Werksseitig hergestellte Zen ent mörtelauskleidungen für Guss-und Stahlrohre-Anforderungen und Prüfungen, Einsatzberelche."

Holtschulte, H., and M.R. Schock. 1985. Asbestos-Cement and Cement-Mortar-Lined Pipes. In *Internal Corrosion of Water Distribution Systems*; Cooperative Research Report, AWWA Research Foundation and DVGW-Forschungsstelle. Denver: AWWARF.

JM Montgomery, Consulting Engineers, Inc. 1985. *Water Treatment Principles and Design*. New York: John Wiley & Sons.

Miller, W.T. 1965. Durability of Cement-Mortar Linings in Cast-Iron Pipe. *Jour. AWWA*, 57:6:773.

Miller, H.C., W.J. Barrett, and R.H. James. 1982. Investigating Potential Water Contamination by Petroleum-Asphalt Coatings in Ductile-Iron Pipe. *Jour. AWWA*, 74:3:151.

National Sanitation Foundation. 1988. Drinking Water System Components, Health Effects. NSF Standard 61. Ann Arbor, MI: NSF.

Radziul, J.V., and R.E. Jackson. 1965. Cleaning and Lining of Small Pipe at Philadelphia. *Jour. AWWA*, 57:12:1513.

Smith, R.L. and C.F. Raba. 1980. The Basic Chemistry of Fly Ash Pozzolanic Reactions. Presented at the Spring Meeting of the Texas Section of American Society of Civil Engineers.

Speller, F.N. 1951. *Corrosion, Causes and Prevention*. 3rd Ed. New York: McGraw-Hill Book Co., Inc.

State of California. 1983. California Test 403: Method of Testing Cement Treated Base and Concrete for Calcium Oxide. Sacramento: Department of Transportation, Division of Construction.

Steel, E.W. 1960. *Water Supply and Sewerage*. 4th Ed. New York: McGraw-Hill Book Co., Inc.

Stevens, A.A., L.A. Moore, and R.J. Miltner. 1989. Formation and Control of Non-Trihalomethane Disinfection By-products. *Jour. AWWA*, 81:8:54.

Troxell, G.E., H.E. Davis, and J.W. Kelly. 1968. *Composition and Properties of Concrete*. 2nd Ed. New York: McGraw-Hill Book Co., Inc.

Uhlig, H.H., ed. 1948. *The Corrosion Handbook*. Sponsored by The Electrochemical Society. New York: John Wiley & Sons.

Warren, I.C., and A.F. Elvidge. 1988. Water Industry Corrosion Protection Lining. Presented at The International Society for Trenchless Technology/Water Pollution Control Federation Conference.

Abbreviations

°C	degrees Celsius
EPA	United States Environmental Protection Agency
°F	degrees Fahrenheit
ft	feet, foot
fps	feet per second
g	gram
gal	gallon
gpm	gallons per minute
in.	inch
Kg/m^3	kilograms per cubic meter
L	liter
lb/ft^3	pounds per cubic foot
lbf	pounds force
lfpm	linear feet per minute
LI	Langlier Index
lmps	linear meters per second
lps	liters per second
m	meter
MCL	maximum contaminant level
mgd	million gallons per day
mg/L	milligrams per liter
mg/mm^2	milligrams per square millimeter
mm	millimeter
mol/m^3	gram–moles per cubic meter
mps	meters per second
MUD	Municipal Utility District
MWD	Municipal Water District
N	Newton
NSF	National Sanitation Foundation
PAH	polynuclear aromatic hydrocarbon
PCE	tetrachloroethylene ("perchloroethylene")
psi	pounds per square inch
PVC	polyvinyl chloride
RWA	Regional Water Authority
SC	Sanitary Commission
SDWA	Safe Drinking Water Act
THM	trihalomethane
μg/L	microgram per liter

Appendix A

Field Test Results

- South Central Connecticut RWA Results
- Marin MWD Results
- Cement Mill Analyses

Page No. 1
09/05/90

AWWARF PROJECT - CORROSION CONTROL OF CEMENT MORTAR LINING

Utility: South Central Connecticut Regional Water Authority

Date Collected	pH	ALK	Ca	Si	Zn	Mg	OPhos	TPhos	Fe	Mn
** Location (34900) INLET WATER (FINISHED WATER)										
04/30/90	7.11	12	14	7.98	0.12	2.0		0.92	0.01	<0.01
05/01/90	7.19	12								
05/02/90	7.23	12	14							
05/03/90	7.19	12								
05/04/90	7.21	12	14							
05/05/90	7.33	12								
05/06/90	7.33	12								
05/07/90	6.96	10	14	7.79	0.14	2.0		1.07	0.01	<0.01
05/08/90	7.27	12								
05/09/90	7.22	11	14							
05/10/90	7.21	13								
05/11/90	7.20	12	15							
05/12/90	7.12	13								
05/13/90	7.28	13								
05/14/90	7.26	12	12	6.85	0.13	1.9		0.95	0.01	<0.01
05/15/90	7.25	13								
05/16/90	7.17	13								
05/17/90	7.02	12								
05/18/90	7.05	13								
05/19/90	7.19	13								
05/20/90	7.11	13								
05/21/90	7.14	13	12			1.9				
05/22/90	7.17	13								
05/23/90	7.25	12								
05/24/90	7.25	12								
05/25/90	7.08	13								
05/26/90	7.30	13								
05/27/90	7.12	12								
05/28/90	7.02	12	10	7.74	0.15	1.8		0.77	0.05	0.02
05/29/90	7.19	13								
05/30/90	7.30	13								
05/31/90	7.38	13								
06/01/90	7.27	13								
06/02/90	7.21	13								
06/03/90	7.26	14								
06/04/90	7.16	13	13							
06/05/90	7.25	14								
06/06/90	7.20	13								
06/07/90	7.18	13								
06/08/90	7.19	13								
06/09/90	7.18	13								
06/10/90	7.18	13								
06/11/90	7.17	14	10							
06/12/90	7.09	13								
06/13/90	7.19	13								
06/14/90	7.09	13								

Page No. 2
09/05/90

AWWARF PROJECT - CORROSION CONTROL OF CEMENT MORTAR LINING

Utility: South Central Connecticut Regional Water Authority

Date Collected	pH	ALK	Ca	Si	Zn	Mg	OPhos	TPhos	Fe	Mn
06/15/90	7.06	13								
06/16/90	7.03	13								
06/17/90	7.17	14								
06/18/90	7.14	13	10							
06/19/90	7.08	14								
06/20/90	7.23	14								
06/21/90	7.10	14								
06/22/90	7.21	13								
06/23/90	7.22	13								
06/24/90	7.31	14								
06/25/90	6.90	13	11	7.39	0.12	1.9		1.32	0.02	0.01
06/26/90	7.32	13								
06/27/90	7.27	14								
06/28/90	7.38	14								
06/29/90	7.34	14								
06/30/90	7.30	14								
07/01/90	7.42	15								
07/02/90	7.20	15	12							
07/03/90	7.16	15								
07/04/90	7.26	15								
07/05/90	7.13	15								
07/06/90	7.23	15								
07/07/90	7.24	15								
07/08/90	7.29	15								
07/09/90	7.31	15	14							
07/10/90	7.24	15								
07/11/90	7.22	15								
07/12/90	7.10	15								
07/13/90	7.25	15								
07/14/90	7.16	15								
07/15/90	7.25	15								
07/16/90	7.14	15	13							
07/17/90	7.16	15								
07/18/90	7.27	15								
07/19/90	7.17	15								
07/20/90	7.15	15								
07/21/90	7.14	15								
07/22/90	7.13	15								
07/23/90	6.94	14	14							
07/24/90	7.22	15								
07/25/90	7.26	15								
07/26/90	7.31	16								
07/27/90	7.23	15								
07/28/90	7.19	15								
07/29/90	7.27	16								
07/30/90	7.08	14	13	7.86	0.11	1.9		1.35	0.02	0.01
07/31/90	7.19	16								

Page No. 3
09/05/90

AWWARF PROJECT - CORROSION CONTROL OF CEMENT MORTAR LINING

Utility: South Central Connecticut Regional Water Authority

Date Collected	pH	ALK	Ca	Si	Zn	Mg	OPhos	TPhos	Fe	Mn
** Location (34901) SECTION 1 (REFERENCE MORTAR, CONTINUOUS FLOW)										
04/30/90	7.11	12	13							
05/01/90	7.19	13								
05/02/90	7.22	13	14							
05/03/90	7.17	12								
05/04/90	7.19	12	15							
05/05/90	7.32	12								
05/06/90	7.26	12								
05/07/90	6.94	11	14	7.88	0.13	2.0		1.13	0.01	<0.01
05/08/90	7.23	12								
05/09/90	7.20	11	14							
05/10/90	7.23	13								
05/11/90	7.18	13	14							
05/12/90	7.15	13								
05/13/90	7.28	12								
05/14/90	7.28	11	12	6.92	0.13	1.9		0.86	0.01	<0.01
05/15/90	7.25	12								
05/16/90	7.12	13								
05/17/90	7.02	12								
05/18/90	7.06	13								
05/19/90	7.15	13								
05/20/90	7.10	12								
05/21/90	7.14	13	12			1.9				
05/22/90	7.17	13								
05/23/90	7.22	12								
05/24/90	7.22	12								
05/25/90	7.09	13								
05/26/90	7.28	13								
05/27/90	7.07	12								
05/28/90	7.00	12	12	7.77	0.14	1.8		0.92	0.03	0.02
05/29/90	7.18	13								
05/30/90	7.30	14								
05/31/90	7.34	13								
06/01/90	7.25	13								
06/02/90	7.19	13								
06/03/90	7.24	14								
06/04/90	7.18	14	13							
06/05/90	7.19	13								
06/06/90	7.17	13								
06/07/90	7.25	13								
06/08/90	7.17	13								
06/09/90	7.17	13								
06/10/90	7.14	13								
06/11/90	7.13	14	10							
06/12/90	7.11	13								
06/13/90	7.17	13								

Page No. 4
09/05/90

AWWARF PROJECT - CORROSION CONTROL OF CEMENT MORTAR LINING

Utility: South Central Connecticut Regional Water Authority

Date Collected	pH	ALK	Ca	Si	Zn	Mg	OPhos	TPhos	Fe	Mn
06/14/90	7.11	13								
06/15/90	7.10	13								
06/16/90	7.07	13								
06/17/90	7.14	14								
06/18/90	7.10	14	10							
06/19/90	7.10	14								
06/20/90	7.22	14								
06/21/90	7.11	14								
06/22/90	7.18	13								
06/23/90	7.20	13								
06/24/90	7.29	15								
06/25/90	6.93	13	12	7.40	0.12	1.9		1.17	0.01	<0.01
06/26/90	7.24	13								
06/27/90	7.28	14								
06/28/90	7.32	14								
06/29/90	7.30	14								
06/30/90	7.32	14								
07/01/90	7.34	15								
07/02/90	7.20	15	12							
07/03/90	7.18	15								
07/04/90	7.28	15								
07/05/90	7.11	15								
07/06/90	7.26	15								
07/07/90	7.26	15								
07/08/90	7.22	15								
07/09/90	7.29	15	14							
07/10/90	7.20	15								
07/11/90	7.20	15								
07/12/90	7.09	15								
07/13/90	7.23	15								
07/14/90	7.19	15								
07/15/90	7.20	15								
07/16/90	7.12	15	14							
07/17/90	7.18	15								
07/18/90	7.24	16								
07/19/90	7.15	15								
07/20/90	7.13	15								
07/21/90	7.12	15								
07/22/90	7.20	16								
07/23/90	6.87	13	14							
07/24/90	7.18	16								
07/25/90	7.23	15								
07/26/90	7.26	15								
07/27/90	7.21	15								
07/28/90	7.21	15								
07/29/90	7.28	16								
07/30/90	7.07	14	14	7.97	0.11	1.9		1.23	0.01	0.01

Page No. 5
09/05/90

AWWARF PROJECT - CORROSION CONTROL OF CEMENT MORTAR LINING

Utility: South Central Connecticut Regional Water Authority

Date Collected	pH	ALK	Ca	Si	Zn	Mg	OPhos	TPhos	Fe	Mn
07/31/90	7.22	15								
** Location (34902) SECTION 2 (REFERENCE MORTAR, DAILY FLUSH)										
05/01/90	11.70	390								
05/02/90	11.81	151	80							
05/03/90	11.75	140								
05/04/90	11.54	109	70							
05/05/90	11.60	111								
05/06/90	11.56	102								
05/07/90	11.17	66	48	8.35	0.07	1.9		0.83	0.01	<0.01
05/08/90	11.53	100								
05/09/90	10.70	53	41							
05/10/90	11.06	73								
05/11/90	10.78	63	50							
05/12/90	11.03	64								
05/13/90	11.08	70								
05/14/90	10.58	56	40	7.74	0.07	1.7		0.64	<0.01	<0.01
05/15/90	10.70	61								
05/16/90	11.05	63								
05/17/90	10.98	54								
05/18/90	11.08	57								
05/19/90	10.90	58								
05/20/90	11.17	67								
05/21/90	10.48	40	28			1.8				
05/22/90	10.94	56								
05/23/90	11.36	63								
05/24/90	10.58	43								
05/25/90	10.84	53								
05/26/90	10.78	56								
05/27/90	10.17	59								
05/28/90	10.32	36	26	8.02	0.10	1.7		0.83	0.01	0.01
05/29/90	10.69	54								
05/30/90	10.84	55								
05/31/90	10.69	51								
06/01/90	10.54	49								
06/02/90	10.78	54								
06/03/90	10.62	56								
06/04/90	10.31	35	30							
06/05/90	10.88	49								
06/06/90	10.91	55								
06/07/90	10.49	41								
06/08/90	10.51	43								
06/09/90	10.41	39								
06/10/90	10.87	59								
06/11/90	10.37	34	26							
06/12/90	10.58	47								
06/13/90	10.61	43								

Page No. 6
09/05/90

AWWARF PROJECT - CORROSION CONTROL OF CEMENT MORTAR LINING

Utility: South Central Connecticut Regional Water Authority

Date Collected	pH	ALK	Ca	Si	Zn	Mg	OPhos	TPhos	Fe	Mn
06/14/90	10.51	41								
06/15/90	10.31	42								
06/16/90	10.40	43								
06/17/90	10.36	46								
06/18/90	10.48	36	26							
06/19/90	10.70	51								
06/20/90	10.47	43								
06/21/90	10.01	34								
06/22/90	10.58	46								
06/23/90	10.34	41								
06/24/90	10.91	62								
06/25/90	10.19	34	26	7.66	0.08	1.8		1.04	0.02	0.01
06/26/90	10.64	50								
06/27/90	10.28	36								
06/28/90	9.64	28								
06/29/90	10.5	41								
06/30/90	10.23	37								
07/01/90	10.59	47								
07/02/90	10.12	34	25							
07/03/90	10.56	45								
07/04/90	10.44	39								
07/05/90	10.43	40								
07/06/90	10.53	47								
07/07/90	10.55	41								
07/08/90	10.49	42								
07/09/90	9.77	31	28							
07/10/90	10.68	49								
07/11/90	10.33	39								
07/12/90	10.05	34								
07/13/90	10.39	40								
07/14/90	10.43	45								
07/15/90	10.39	43								
07/16/90	9.75	31	28							
07/17/90	10.39	43								
07/18/90	10.47	43								
07/19/90	9.85	35								
07/20/90	10.36	41								
07/21/90	10.19	39								
07/22/90	10.44	41								
07/23/90	9.97	32	29							
07/24/90	10.44	45								
07/25/90	9.99	37								
07/26/90	10.15	41								
07/27/90	10.15	39								
07/28/90	9.98	38								
07/29/90	10.46	43								
07/30/90	9.85	31	27	8.36	0.09	1.9		1.10	<0.01	0.01

Page No. 7
09/05/90

AWWARF PROJECT - CORROSION CONTROL OF CEMENT MORTAR LINING

Utility: South Central Connecticut Regional Water Authority

Date Collected	pH	ALK	Ca	Si	Zn	Mg	OPhos	TPhos	Fe	Mn
07/31/90	10.38	43								
** Location (34903)			SECTION 3		(REFERENCE MORTAR, WEEKLY FLUSH)					
05/07/90	12.54	756	135	25.92	0.07	1.1		0.43	0.01	0.01
05/14/90	11.68	330	124	18.60	0.02	0.4		0.21	0.01	<0.01
05/21/90	11.69	200	82			1.0				
05/28/90	11.48	156	64	9.84	0.01	0.8		0.34	0.01	<0.01
06/04/90	11.50	128	51							
06/11/90	11.45	108	38							
06/18/90	11.43	93	36							
06/25/90	11.25	83	35	7.19	0.03	1.6		0.86	0.04	0.01
07/02/90	11.12	73	32							
07/09/90	11.03	70	35							
07/16/90	10.89	70	36							
07/23/90	10.98	78	38							
07/30/90	10.90	65	37	7.94	0.04	1.9		1.96	0.02	0.01
** Location (34904)			SECTION 4		(CENTRIFUGALLY-LINED, NO SEAL COAT)					
05/07/90	11.86	284	29	95.93	0.08	0.2		0.18	1.76	<0.01
05/14/90	11.28	170	108	47.74	0.03	0.2		0.09	0.30	<0.01
05/21/90	11.58	176	127			0.1				
05/28/90	11.49	172	141	34.35	0.02	0.1		<0.03	0.15	0.01
06/04/90	11.59	172	149							
06/11/90	11.57	156	141							
06/18/90	11.68	141	130							
06/25/90	11.53	136	116	23.56	0.04	0.2		0.21	0.17	0.01
07/02/90	11.46	128	120							
07/09/90	11.39	132	110							
07/16/90	11.26	117	112							
07/23/90	11.31	111	107							
07/30/90	11.11	82	78	13.08	0.05	0.8		0.52	0.04	0.01
** Location (34905)			SECTION 5		(CENTRIFUGALLY-LINED, SEAL COAT)					
05/07/90	11.84	204	30	14.08	0.01	0.3		0.37	0.37	<0.01
05/14/90	10.91	80	25	7.84	0.07	1.7		0.61	0.11	<0.01
05/21/90	10.97	60	20			1.8				
05/28/90	10.71	52	20	7.26	0.05	1.8		0.77	0.14	0.01
06/04/90	10.63	41	21							
06/11/90	10.56	37	18							
06/18/90	10.40	34	15							
06/25/90	10.23	33	18	6.89	0.01	1.9		2.45	0.06	0.01
07/02/90	10.07	31	18							
07/09/90	10.00	32	18							
07/16/90	9.82	31	21							
07/23/90	9.97	29	21							
07/30/90	9.83	29	22	8.20	0.05	2.0		1.99	0.03	0.01

Page No. 8
09/05/90

AWWARF PROJECT - CORROSION CONTROL OF CEMENT MORTAR LINING

Utility: South Central Connecticut Regional Water Authority

Date Collected	pH	ALK	Ca	Si	Zn	Mg	OPhos	TPhos	Fe	Mn
** Location (34906)			SECTION 6		(REFERENCE MORTAR, CONDITIONED WATER)					
05/07/90	12.62	916	292	11.78	0.02	0.2			0.03	<0.01
05/14/90	11.75	422	252	23.76	0.07	0.3			0.02	<0.01
05/21/90	11.93	324	190			0.6				
05/28/90	11.81	268	186	22.59	0.07	0.4			0.02	0.01
06/04/90	11.84	248	185							
06/11/90	11.79	204	123							
06/18/90	11.86	180	103							
06/25/90	11.65	163	116	16.97	0.03	0.3			0.02	<0.01
07/02/90	11.58	145	132							
07/09/90	11.50	132	120							
07/16/90	11.35	124	107							
07/23/90	11.38	117	111							
07/30/90	11.35	110	97	11.22	0.02	0.5			<0.01	<0.01
** Location (34907)			SECTION 7		(HIGH SAND MORTAR)					
05/07/90	12.45	692	138	30.11	0.02	0.5		0.37	0.01	<0.01
05/14/90	11.65	320	214	27.67	0.02	0.3		0.12	0.01	<0.01
05/21/90	11.77	244	168			0.2				
05/28/90	11.61	208	168	25.35	0.01	0.1		0.25	0.02	<0.01
06/04/90	11.68	188	151							
06/11/90	11.63	152	90							
06/18/90	11.67	135	85							
06/25/90	11.48	122	101	16.84	0.04	0.6		0.18	0.07	0.02
07/02/90	11.39	110	92							
07/09/90	11.29	108	86							
07/16/90	11.14	95	88							
07/23/90	11.25	90	82							
07/30/90	11.18	86	80	9.37	0.02	1.2		0.52	0.03	0.01
** Location (34908)			SECTION 8		(TYPE II CEMENT MORTAR)					
05/07/90	12.41	676	63	59.64	0.02	0.3		0.28	0.18	<0.01
05/14/90	11.67	330	196	35.14	0.03	0.3		0.06	0.05	<0.01
05/21/90	11.87	296	215			0.2				
05/28/90	11.76	264	175	32.97	0.02	0.1		<0.03	0.02	0.01
06/04/90	11.82	260	194							
06/11/90	11.79	220	141							
06/18/90	11.86	198	108							
06/25/90	11.65	180	127	27.09	0.03	0.3		0.21	0.05	0.01
07/02/90	11.58	157	124							
07/09/90	11.48	148	103							
07/16/90	11.31	130	102							
07/23/90	11.34	118	94							
07/30/90	11.32	108	84	15.26	0.02	1.0		0.37	0.02	0.01
** Location (34909)			SECTION 9		(WATER-REDUCING ADMIXTURE MORTAR)					
05/07/90	12.46	740	54	49.48	0.02	0.3		0.21	0.07	<0.01

Page No. 9
09/05/90

AWWARF PROJECT - CORROSION CONTROL OF CEMENT MORTAR LINING

Utility: South Central Connecticut Regional Water Authority

Date Collected	pH	ALK	Ca	Si	Zn	Mg	OPhos	TPhos	Fe	Mn
05/14/90	11.70	358	182	33.00	0.02	0.3		0.15	0.03	<0.01
05/21/90	11.90	304	171			0.2				
05/28/90	11.79	272	186	29.42	0.01	0.1		<0.03	0.01	0.01
06/04/90	11.83	260	149							
06/11/90	11.81	212	133							
06/18/90	11.85	186	90							
06/25/90	11.63	165	120	20.15	0.03	0.3		0.28	0.18	0.02
07/02/90	11.53	138	102							
07/09/90	11.40	124	84							
07/16/90	11.21	116	78							
07/23/90	11.23	96	70							
07/30/90	11.20	86	62	9.33	0.02	1.4		1.35	0.02	0.01
** Location (34910)			SECTION 10		(POZZOLAN MORTAR)					
05/07/90	12.56	788	78	41.47	0.03	0.3		0.21	0.02	<0.01
05/14/90	11.70	352	148	29.72	0.02	0.3		0.37	<0.01	<0.01
05/21/90	11.82	260	135			0.2				
05/28/90	11.67	192	95	19.39	0.01	0.2		<0.03	<0.01	<0.01
06/04/90	11.60	168	99							
06/11/90	11.62	140	91							
06/18/90	11.56	114	60							
06/25/90	11.39	107	66	9.54	0.03	1.1		0.52	0.04	0.01
07/02/90	11.29	95	63							
07/09/90	11.16	84	50							
07/16/90	11.06	85	63							
07/23/90	11.09	81	60							
07/30/90	11.07	76	57	9.06	0.03	1.5		1.17	0.02	0.01
** Location (34911)			CONDITIONED WATER							
04/30/90	6.72	9	14	7.59	0.30	1.9	0.09		0.02	<0.01
05/07/90	6.98	9	14	7.55	0.22	2.0	0.06		0.01	<0.01
05/14/90	6.39	8	12	7.08	0.23	1.9	0.06		0.02	<0.01
05/21/90	6.86	10	12			2.0				
05/28/90	6.70	10	12	7.03	0.32	1.8			0.05	0.01
06/04/90	6.85	10	13							
06/11/90	6.88	14	12							
06/18/90	6.85	10	10							
06/25/90	6.72	10	11	6.92	0.19	1.8	0.09		<0.01	<0.01
07/02/90	6.78	10	12							
07/09/90	6.88	9	14							
07/16/90	6.87	10	14							
07/23/90	6.38	9	13							
07/30/90	6.92	9	14	8.03	0.19	2.0	0.18		<0.01	<0.01

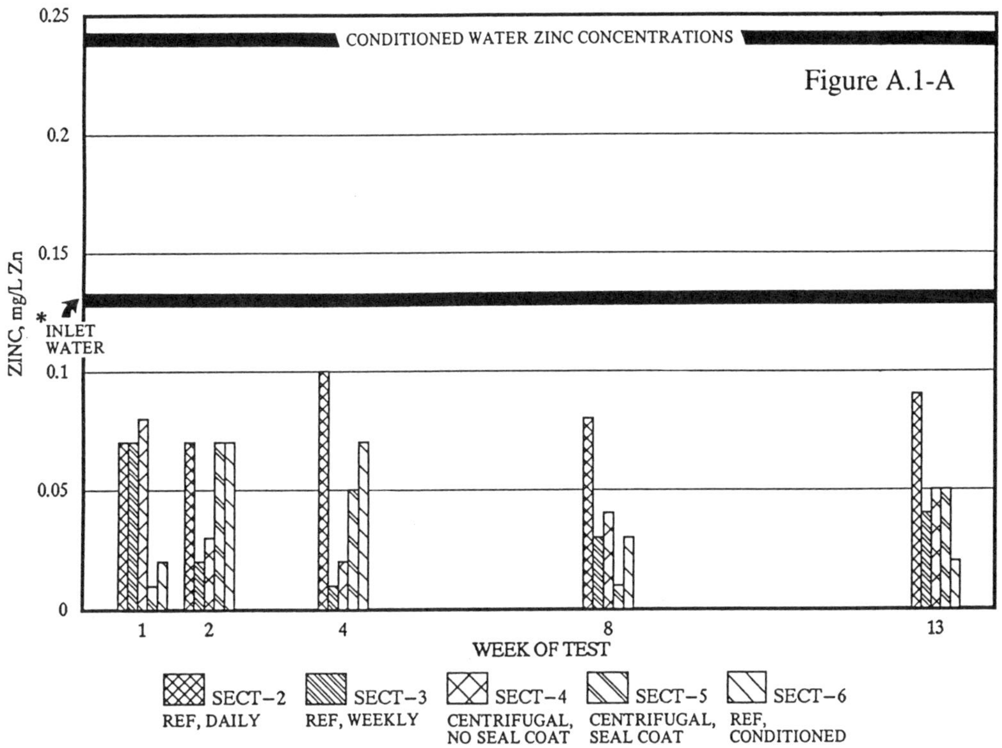

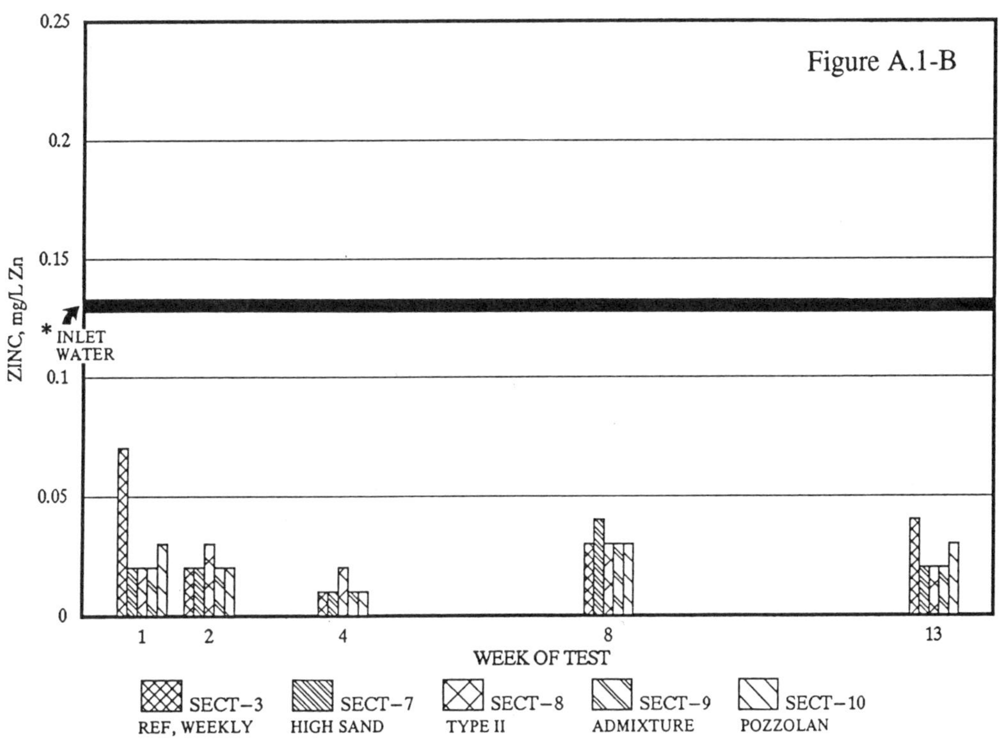

*SECT–1 (Reference Mortar, Continuous Flow) had the same composition as the Inlet Water.

Figure A.1 South Central Connecticut RWA zinc results

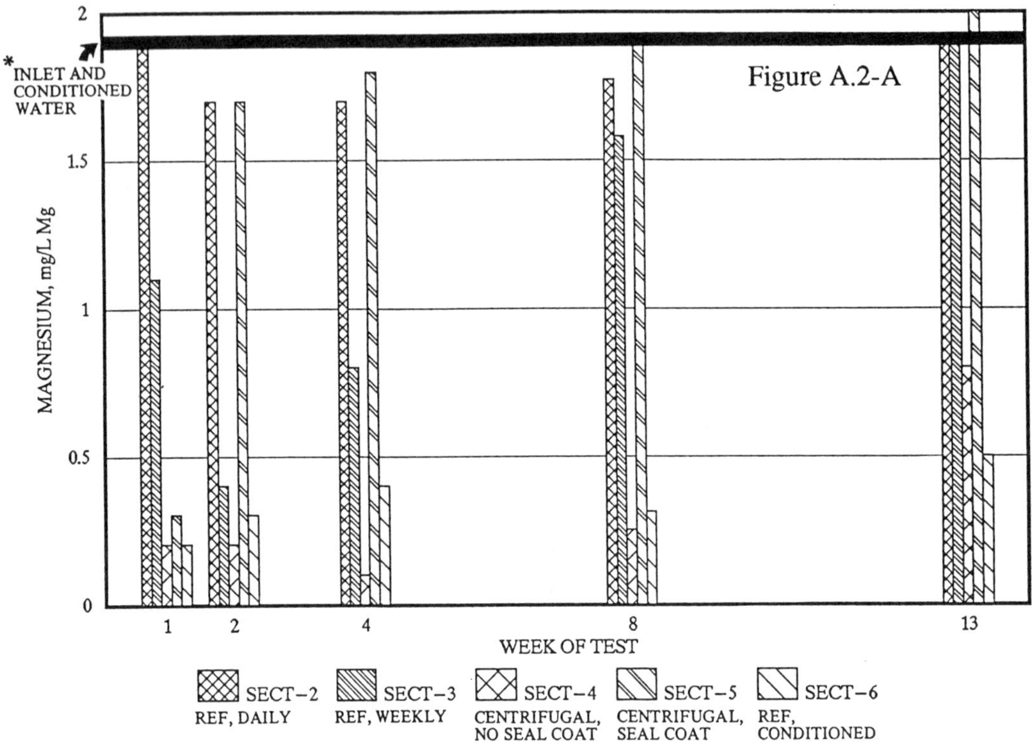

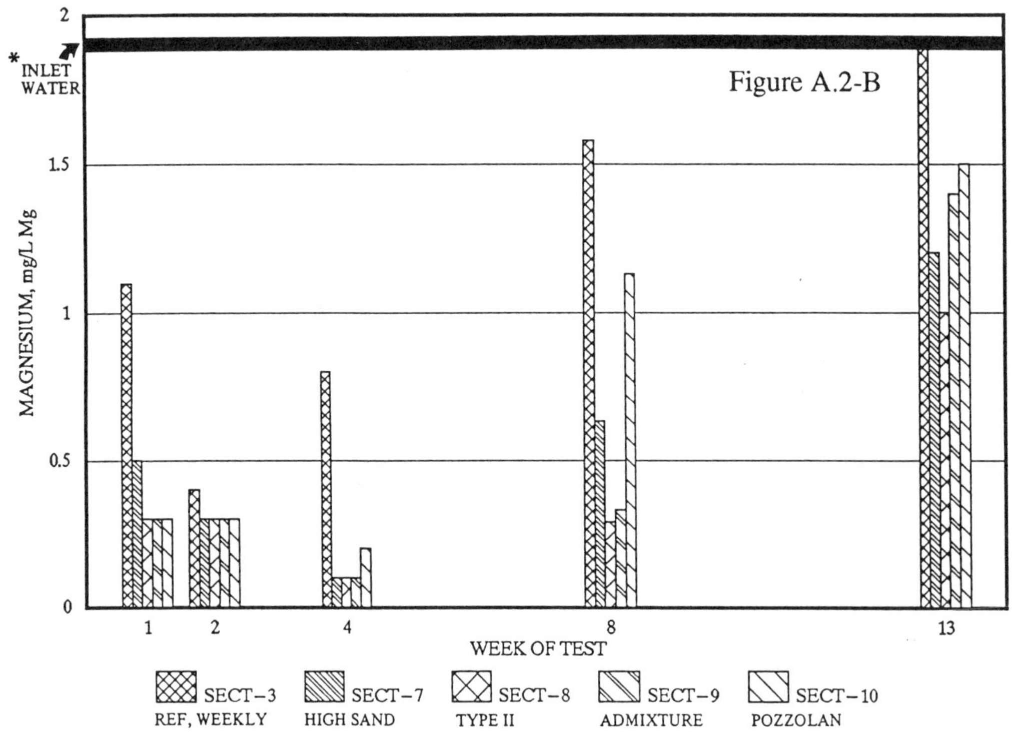

*SECT−1 (Reference Mortar, Continuous Flow) had the same composition as the Inlet Water.

Figure A.2 South Central Connecticut RWA magnesium results

Utility: Marin Municipal Water District

INLET WATER

DATE	pH	ALK	Ca	Si	Zn	Mg	OPhos	TPhos	Fe	Mn
MAY 1	7.7	71	42							
MAY 2	7.7	71	41							
MAY 3	7.9									
MAY 4	7.75	72	39							
MAY 5	7.8									
MAY 6	7.77									
MAY 7	7.7	72	41	3.4	0.32	9	1.08		<0.05	<0.03
MAY 8	7.8									
MAY 9	7.68	74	44							
MAY 10	7.75									
MAY 11	7.7	72	40							
MAY 12	7.87									
MAY 13	8.2									
MAY 14	7.6	72	38	2.9	0.37	10	0.92		<0.05	<0.03
MAY 15	7.95									
MAY 16	7.84	72	38							
MAY 17	7.87									
MAY 18	7.9									
MAY 19	7.8									
MAY 20	8.0									
MAY 21	7.96	74	37	9	0.32	8.4	0.9		<0.05	<0.03
MAY 22	7.82									
MAY 23	7.9	74	36							
MAY 24	7.84									
MAY 25	7.87	72	36							
MAY 26	7.76									
MAY 27	7.68									
MAY 28	7.87	72	32	8.7	0.39	9.3	1.04		<0.05	<0.03
MAY 29	7.8									
MAY 30	7.85	72	36							
MAY 31	7.87									
JUNE 1	7.9									
JUNE 2	8									
JUNE 3	7.8									
JUNE 4	7.7	70	36							
JUNE 5	7.8									
JUNE 6	7.89									
JUNE 7	7.9									
JUNE 8	7.8									
JUNE 9	7.8									
JUNE 10	7.81									
JUNE 11	8.3	73	28							
JUNE 12	7.87									
JUNE 13	7.9									
JUNE 14	7.9									
JUNE 15	7.9									

Utility: Marin Municipal Water District

DATE	pH	ALK	Ca	Si	Zn	Mg	OPhos	TPhos	Fe	Mn
JUNE 16	7.81									
JUNE 17	7.87									
JUNE 18	7.82	71	38							
JUNE 19	8.04									
JUNE 20	7.76									
JUNE 21	7.88									
JUNE 22	7.82									
JUNE 23	7.9									
JUNE 24	7.6									
JUNE 25	7	55	39	3.8	0.52	9.5	0.96		<0.05	<0.03
JUNE 26	7.66									
JUNE 27	7.68									
JUNE 28	7.76									
JUNE 29	7.88									
JUNE 30	7.82									
JULY 1	7.82									
JULY 2	8.02	74	37							
JULY 3	7.82									
JULY 4	7.86									
JULY 5	7.8									
JULY 6	7.3									
JULY 7	6.8									
JULY 8	6.78									
JULY 9	6.9	54	16							
JULY 10	6.9									
JULY 11										
JULY 12	6.8									
JULY 13	6.7									
JULY 14	6.8									
JULY 15	7.09									
JULY 16	6.94	57	42							
JULY 17	6.91									
JULY 18	6.9									
JULY 19	6.9									
JULY 20	7									
JULY 21	6.93									
JULY 22	6.75									
JULY 23	6.89	55	34							
JULY 24	6.72									
JULY 25	6.85									
JULY 26	6.87									
JULY 27	6.82									
JULY 28	6.88									
JULY 29	6.77									
JULY 30	6.72	54	39	6	0.38	10.2	0.87		<0.05	<0.03

Utility: Marin Municipal Water District

SECTION 1: REFERENCE MORTAR, CONTINUOUS FLOW

DATE	pH	ALK	Ca	Si	Zn	Mg	OPhos	TPhos	Fe	Mn
MAY 1	7.8									
MAY 2	7.8	71	40							
MAY 3	7.7									
MAY 4	7.8	72	39							
MAY 5	7.8									
MAY 6	7.67									
MAY 7	7.7	72	42	3.5	0.32	9	0.94		<0.05	<0.03
MAY 8	7.67									
MAY 9	8	74	45							
MAY 10	7.95									
MAY 11	7.7	72	42							
MAY 12	7.9									
MAY 13	7.92									
MAY 14	7.6	74	58	3.1	0.38	5	1.02		<0.05	<0.03
MAY 15	7.8									
MAY 16	7.8	73	41							
MAY 17	7.8									
MAY 18	7.8									
MAY 19	7.8									
MAY 20	7.95									
MAY 21	7.87	74	30	8.8	0.31	10	0.84		<0.05	<0.03
MAY 22	7.82									
MAY 23	7.88	75	35							
MAY 24	7.86									
MAY 25	7.88	72	36							
MAY 26	7.86									
MAY 27	7.65									
MAY 28	7.88	72	37	8.5	0.39	9	1.02		<0.05	<0.03
MAY 29	7.82									
MAY 30	7.85	74	37							
MAY 31	7.8									
JUNE 1	7.9									
JUNE 2	8									
JUNE 3	7.8									
JUNE 4	7.6	70	34							
JUNE 5	7.7									
JUNE 6	7.9									
JUNE 7	7.9									
JUNE 8	7.8									
JUNE 9	7.8									
JUNE 10	7.83									
JUNE 11	8.1	72	36							

Utility: Marin Municipal Water District

DATE	pH	ALK	Ca	Si	Zn	Mg	OPhos	TPhos	Fe	Mn
JUNE 12	7.87									
JUNE 13	8									
JUNE 14	7.9									
JUNE 15	7.9									
JUNE 16	7.93									
JUNE 17	8									
JUNE 18	7.87	70	40							
JUNE 19	8									
JUNE 20	7.78									
JUNE 21	8									
JUNE 22	7.93									
JUNE 23	7.9									
JUNE 24	7.6									
JUNE 25	6.85	54	40	3.7	0.41	9.5	1		<0.05	<0.03
JUNE 26	7.68									
JUNE 27	7.7									
JUNE 28	7.7									
JUNE 29	7.9									
JUNE 30	7.8									
JULY 1	7.79									
JULY 2	7.96	74	39							
JULY 3	7.81									
JULY 4	7.80									
JULY 5	7.7									
JULY 6	7.4									
JULY 7	6.9									
JULY 8	6.83									
JULY 9	7	56	15							
JULY 10	6.9									
JULY 11										
JULY 12	6.8									
JULY 13	6.6									
JULY 14	6.9									
JULY 15	6.97									
JULY 16	6.91	56	40							
JULY 17	6.92									
JULY 18	6.92									
JULY 19	6.9									
JULY 20	7									
JULY 21	7									
JULY 22	6.78									
JULY 23	7	56	39							
JULY 24	6.72									
JULY 25	6.86									
JULY 26	6.81									
JULY 27	6.84									
JULY 28	6.81									
JULY 29	6.79									

Utility: Marin Municipal Water District

DATE	pH	ALK	Ca	Si	Zn	Mg	OPhos	TPhos	Fe	Mn
JULY 30	6.81	58	40	6	0.33	9.7	0.81		0.36	<0.03

SECTION 2: REFERENCE MORTAR, DAILY FLUSH

DATE	pH	ALK	Ca	Si	Zn	Mg	OPhos	TPhos	Fe	Mn
MAY 1										
MAY 2	11.2	157	64							
MAY 3	10.8									
MAY 4	9.5	73	41							
MAY 5	9.5									
MAY 6	10.1									
MAY 7	10.62	132	69	3.6	0.16	3.2	0.82		<0.05	<0.03
MAY 8	10									
MAY 9	10.5	133	59							
MAY 10	10									
MAY 11	10.5	74	41							
MAY 12	8									
MAY 13	10.04									
MAY 14	10.1	122	61	3.3	0.24	9.2	0.9		<0.05	<0.03
MAY 15	10.1									
MAY 16	10.2	120	62							
MAY 17	10									
MAY 18	10									
MAY 19	10.1									
MAY 20	10.12									
MAY 21	10.04	118	60	7.9	0.2	8.7	0.84		<0.05	<0.03
MAY 22	10									
MAY 23	10.1	122	61							
MAY 24	10.01									
MAY 25	10.1	119	66							
MAY 26	10.2									
MAY 27	10.07									
MAY 28	10.17	117	54	8.4	0.47	9.7	1.18		0.05	<0.03
MAY 29	10									
MAY 30	10.03	111	63							
MAY 31	10									
JUNE 1	10									
JUNE 2	9.65									
JUNE 3	9.7									
JUNE 4	9.6	98	53							
JUNE 5	9.8									
JUNE 6	9.2									
JUNE 7	9.9									
JUNE 8	9.9									
JUNE 9	10.1									
JUNE 10	9.83									
JUNE 11	10	113	60							
JUNE 12	9.92									

Utility: Marin Municipal Water District

DATE	pH	ALK	Ca	Si	Zn	Mg	OPhos	TPhos	Fe	Mn
JUNE 13	10									
JUNE 14	10									
JUNE 15	10									
JUNE 16	10									
JUNE 17	10									
JUNE 18	10	104	65							
JUNE 19	10.03									
JUNE 20	10.2									
JUNE 21	10.06									
JUNE 22	10.01									
JUNE 23	9.8									
JUNE 24	9.7									
JUNE 25	9.99	99	63	4.4	0.14	9.2	0.9		<0.05	<0.03
JUNE 26	9.37									
JUNE 27	10									
JUNE 28	10									
JUNE 29	10.1									
JUNE 30	9.66									
JULY 1	9.87									
JULY 2	9.87	104	59							
JULY 3	9.82									
JULY 4	9.76									
JULY 5	9.4									
JULY 6	9.8									
JULY 7	9.3									
JULY 8	9.67									
JULY 9	9.9	106	65							
JULY 10	8.6									
JULY 11										
JULY 12	9.7									
JULY 13	9.8									
JULY 14	8.7									
JULY 15	9.62									
JULY 16	9.67	105	69							
JULY 17	9.67									
JULY 18	9.46									
JULY 19	9.3									
JULY 20	8.5									
JULY 21	9.3									
JULY 22	9.51									
JULY 23	9.8	105	72							
JULY 24	9.63									
JULY 25	9.64									
JULY 26	9.66									
JULY 27	9.71									
JULY 28	9.68									
JULY 29	9.68									
JULY 30	9.37	86	60	6.5	0.46	10	1.4		0.36	<0.03

Utility: Marin Municipal Water District

DATE	pH	ALK	Ca	Si	Zn	Mg	OPhos	TPhos	Fe	Mn
SECTION 3: REFERENCE MORTAR, WEEKLY FLUSH										
MAY 7	11.51	184	68	3.7	0.11	2.9	0.36		<0.05	<0.03
MAY 14	11.3	198	49	2.3	0.02	2.9	0.2		<0.05	<0.03
MAY 21	11.02	170	47	2	0.03	10.2	0.3		<0.05	<0.03
MAY 28	11.02	151	44	4.5	<0.05	8.3	0.52		<0.05	<0.03
JUNE 4	10.9	148	50							
JUNE 11	11	142	58							
JUNE 18	10	100	54							
JUNE 25	11.14	151	66	2.3	<0.05	7.8	0.28		<0.05	<0.03
JULY 2	10.79	132	65							
JULY 9	10.8	134	56							
JULY 16	10.59	135	67							
JULY 23	10.58	130	65							
JULY 30	10.67	126	71	5.3	<0.05	9.7	0.63		<0.05	<0.03
SECTION 4: CENTRIFUGALLY-LINED, NO SEAL COAT										
MAY 7	11.61	251	48	26.9	0.12	2.9	0.36		0.49	<0.03
MAY 14	10.97	153	79	5.4	0.06	6.1	0.26		0.07	<0.03
MAY 21	10.15	122	67	3.3	0.05	10	0.89		<0.05	<0.03
MAY 28	10.13	117	73	7.8	<0.05	9	1.02		<0.05	<0.03
JUNE 4	10	115	68							
JUNE 11	10	115	72							
JUNE 18	10.02	111	70							
JUNE 25	10.1	111	73	4.7	<0.05	10	0.86		<0.05	<0.03
JULY 2	9.87	105	83							
JULY 9	9.9	107	74							
JULY 16	9.66	114	85							
JULY 23	9.87	109	90							
JULY 30	9.82	107	89	5.9	<0.05	9	0.75		<0.05	<0.03
SECTION 5: CENTRIFUGALLY-LINED, SEAL COAT										
MAY 7	10.89	157	40	4.9	0.11	8.7	0.72		0.45	<0.03
MAY 14	9.87	110	44	4.1	0.11	9	0.84		0.43	<0.03
MAY 21	9.61	97	41	3.2	0.08	9.8	0.76		<0.1	<0.03
MAY 28	9.5	94	39	8.3	0.07	9	0.78		0.05	<0.03
JUNE 4	9.4	89	57							
JUNE 11	9.4	86	36							
JUNE 18	9.3	85	37							
JUNE 25	9.28	82	39	4.4	0.09	9.7	0.8		0.05	<0.03
JULY 2	8.8	72	41							
JULY 9	9.3	81	39							

Utility: Marin Municipal Water District

DATE	pH	ALK	Ca	Si	Zn	Mg	OPhos	TPhos	Fe	Mn
JULY 16	8.7	70	41							
JULY 23	8.78	70	41							
JULY 30	8.6	66	41	5.9	0.2	9.7	0.68		<0.05	<0.03
SECTION 6: REFERENCE MORTAR, CONDITIONED WATER										
MAY 7	11.62	267	100	3.2	<0.05	4.4	9.8	21.2	<0.05	<0.03
MAY 14	11.13	196	69	2.9	<0.05	7.3	1.19	2.76	<0.05	<0.03
MAY 21	11.07	170	70	3	<0.05	10	0.32	1.08	<0.05	<0.03
MAY 28	11.12	155	76	2.8	<0.05	8.9	0.26	0.8	<0.05	<0.03
JUNE 4	11	144	72							
JUNE 11	11.1	138	69							
JUNE 18	10.89	125	65							
JUNE 25	11.04	128	65	3.1	<0.05	9.5	0.32	0.52	<0.05	<0.03
JULY 2	10.77	122	61							
JULY 9	10.8	114	62							
JULY 16	10.43	120	64							
JULY 23	10.47	112	60							
JULY 30	10.63	103	62	4.7	<0.05	8.3	0.86	1.12	<0.05	<0.03
SECTION 7: HIGH SAND MORTAR										
MAY 7	11.99	617	380	5.6	0.26	<1	0.66		<0.05	<0.03
MAY 14	11.4	245	120	3.8	0.12	4	0.38		<0.05	<0.03
MAY 21	11.28	147	51	2	0.06	4.9	0.14		<0.05	<0.03
MAY 28	11.13	129	45	3.8	0.09	6.9	0.42		<0.05	<0.03
JUNE 4	10.8	124	52							
JUNE 11	11	123	57							
JUNE 18	10.78	123	56							
JUNE 25	11.06	120	57	2.6	<0.05	8.5	0.44		<0.05	<0.03
JULY 2	10.79	120	71							
JULY 9	10.8	120	62							
JULY 16	10.56	118	80							
JULY 23	10.56	118	80							
JULY 30	10.64	116	84	5.1	<0.05	9	0.54		<0.05	<0.03
SECTION 8: TYPE II CEMENT MORTAR										
MAY 7	11.79	280	70	3.8	0.1	<1.0	0.36		<0.05	<0.03
MAY 14	11.18	182	53	2.5	0.04	7.5	0.26		<0.05	<0.03
MAY 21	10.98	165	53	2.3	0.06	10.2	0.38		<0.05	<0.03
MAY 28	10.94	152	58	6.1	0.06	6.3	0.6		<0.05	<0.03
JUNE 4	10.7	145	61							
JUNE 11	10.9	146	69							
JUNE 18	10.78	142	69							

Utility: Marin Municipal Water District

DATE	pH	ALK	Ca	Si	Zn	Mg	OPhos	TPhos	Fe	Mn
JUNE 25	11.05	147	71	3.8	<0.05	9.5	0.34		<0.05	<0.03
JULY 2	10.83	142	91							
JULY 9	10.8	143	81							
JULY 16	10.63	146	96							
JULY 23	10.63	140	90							
JULY 30	10.7	138	95	5.6	<0.05	9	0.45		<0.05	<0.03

SECTION 9: WATER-REDUCING ADMIXTURE MORTAR

DATE	pH	ALK	Ca	Si	Zn	Mg	OPhos	TPhos	Fe	Mn
MAY 7	11.82	284	100	5.4	0.13	2.4	0.32		<0.05	<0.03
MAY 14	11.2	181	55	2.9	0.08	6.3	0.26		<0.05	<0.03
MAY 21	10.95	161	54	2.6	0.06	10.7	0.52		<0.05	<0.03
MAY 28	10.95	151	61	6.8	<0.05	8.3	0.66		<0.05	<0.03
JUNE 4	10.7	148	60							
JUNE 11	10.9	151	67							
JUNE 18	10.73	144	67							
JUNE 25	11.15	134	69	3	<0.05	6.8	0.46		<0.05	<0.03
JULY 2	10.93	150	61							
JULY 9	10.7	132	59							
JULY 16	10.57	140	84							
JULY 23	10.5	134	83							
JULY 30	10.7	134	87	5.8	<0.05	10	0.5		<0.05	<0.03

SECTION 10: POZZOLAN MORTAR

DATE	pH	ALK	Ca	Si	Zn	Mg	OPhos	TPhos	Fe	Mn
MAY 7	12.04	571	300	6.9	0.19	2.4	0.5		<0.05	<0.03
MAY 14	11.44	251	79	3.5	0.14	1.2	0.36		<0.05	<0.03
MAY 21	11.21	163	54	2	0.1	7.2	0.2		<0.05	<0.03
MAY 28	11.1	145	49	4.5	<0.05	8.5	0.46		<0.05	<0.03
JUNE 4	10.9	138	57							
JUNE 11	11	138	61							
JUNE 18	10.81	134	63							
JUNE 25	11.04	136	60	3	<0.05	9	0.44		<0.05	<0.03
JULY 2	10.83	128	72							
JULY 9	10.8	130	65							
JULY 16	10.59	132	80							
JULY 23	10.55	128	79							
JULY 30	10.67	126	84	5.4	<0.05	9.5	0.58		<0.05	<0.03

CONDITIONED WATER

DATE	pH	ALK	Ca	Si	Zn	Mg	OPhos	TPhos	Fe	Mn
MAY 7	7.61	57	39	3.1	<0.05	9.7	0.24	1	<0.05	<0.03
MAY 14	7.68	58	39	3.1	<0.05	9.0	0.2	0.92	<0.05	<0.03
MAY 21	8.08	57	38	2.9	<0.05	9.5	0.16	0.88	<0.05	<0.03
MAY 28	8.02	57	41	2.9	<0.05	9.5	0.52	0.76	<0.05	<0.03
JUNE 4	8.0	66	38							
JUNE 11	8.3	57	39							
JUNE 18	8.12	39	40							
JUNE 25	8.77	57	42	3.1	<0.05	8.5	0.4	0.4	<0.05	<0.03
JULY 2	8.75	57	38							
JULY 9	8.1	56	40							
JULY 16	8.1	58	41							
JULY 23	8.18	56	40							
JULY 30	8.35	58	59	4.9	<0.05	4.8	0.39	0.48	<0.05	<0.03

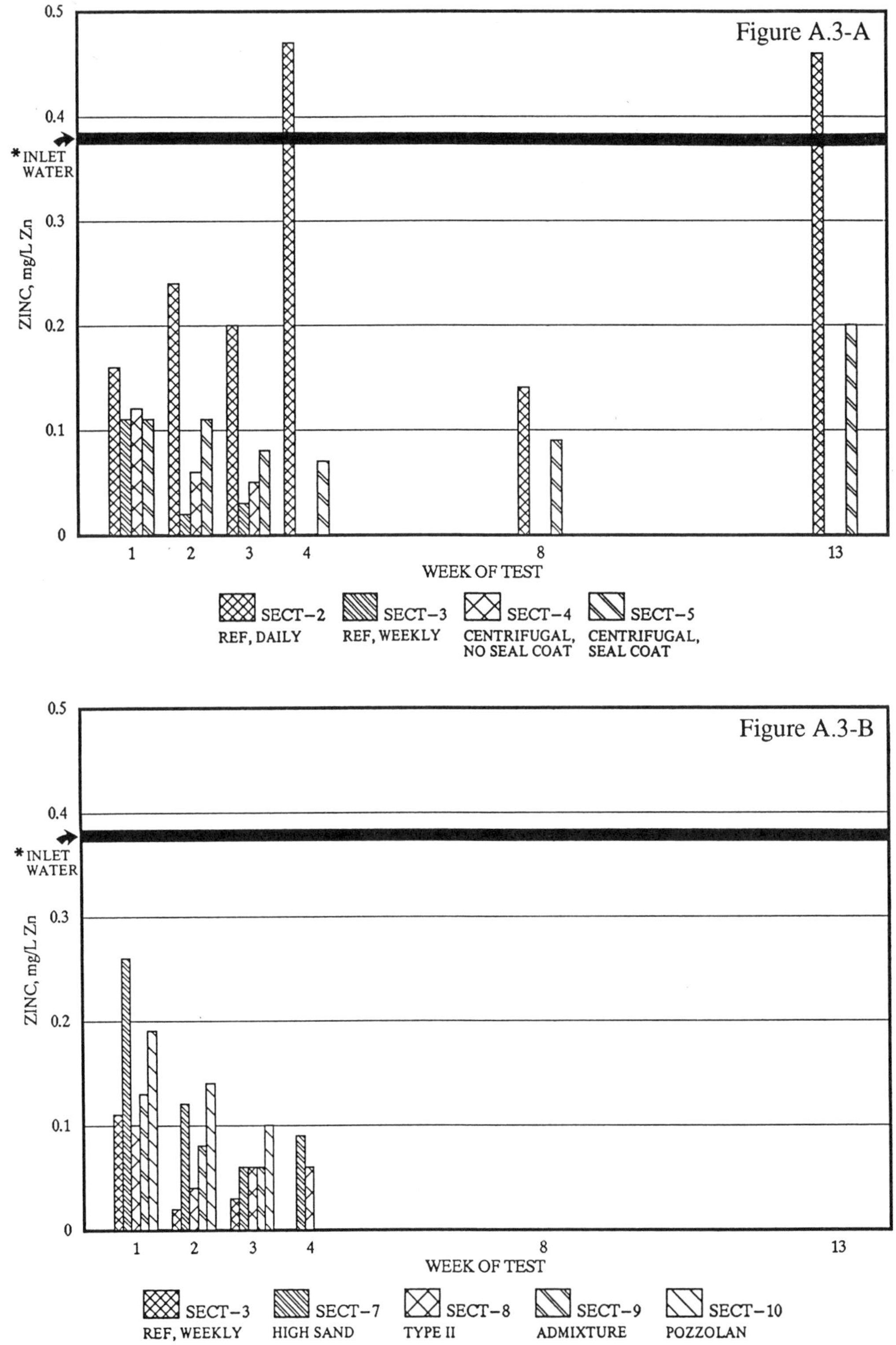

*SECT–1 (Reference Mortar, Continuous Flow) had the same composition as the Inlet Water.

Figure A.3 Marin MWD zinc results

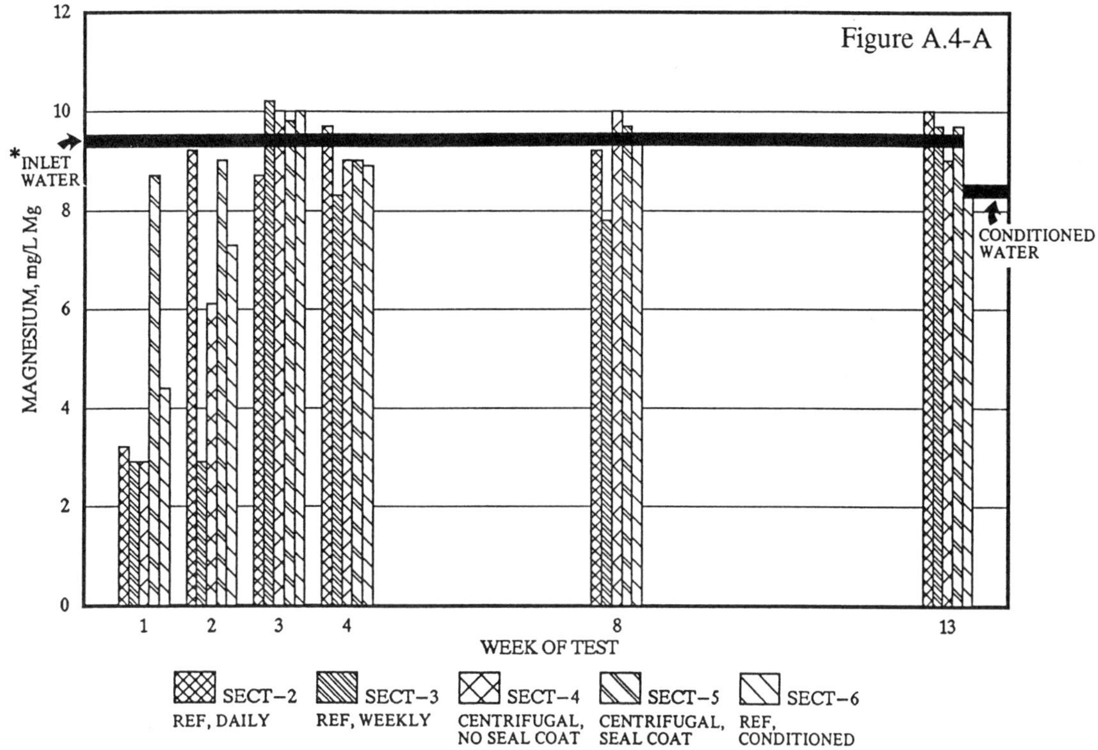

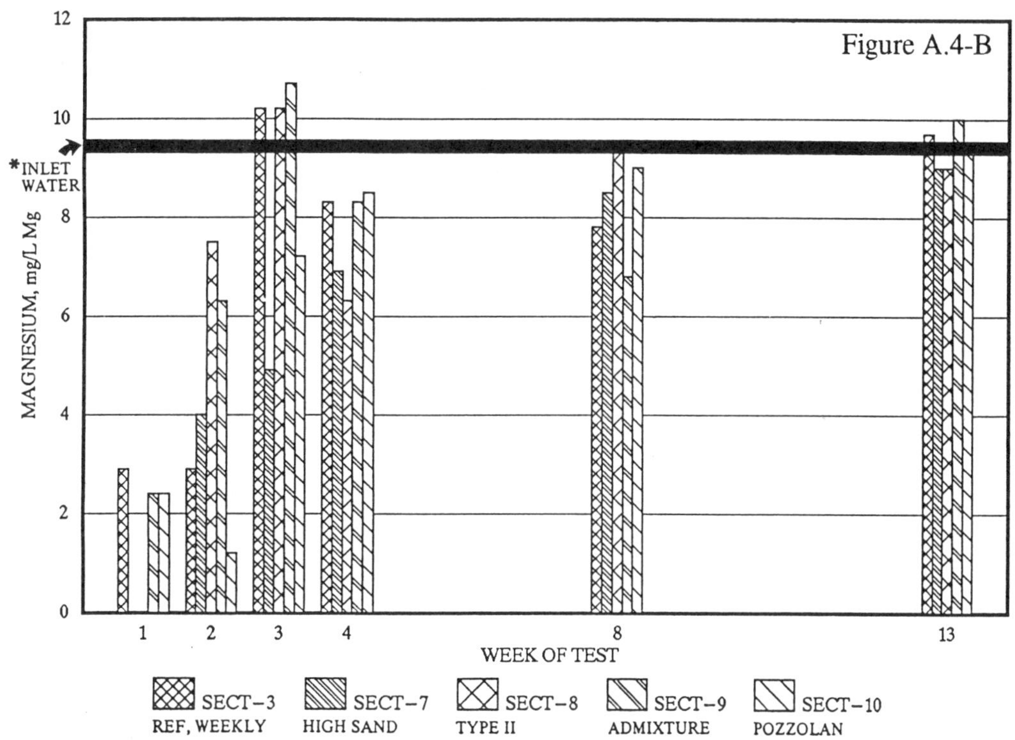

*SECT–1 (Reference Mortar, Continuous Flow) had the same composition as the Inlet Water.

Figure A.4 Marin MWD magnesium results

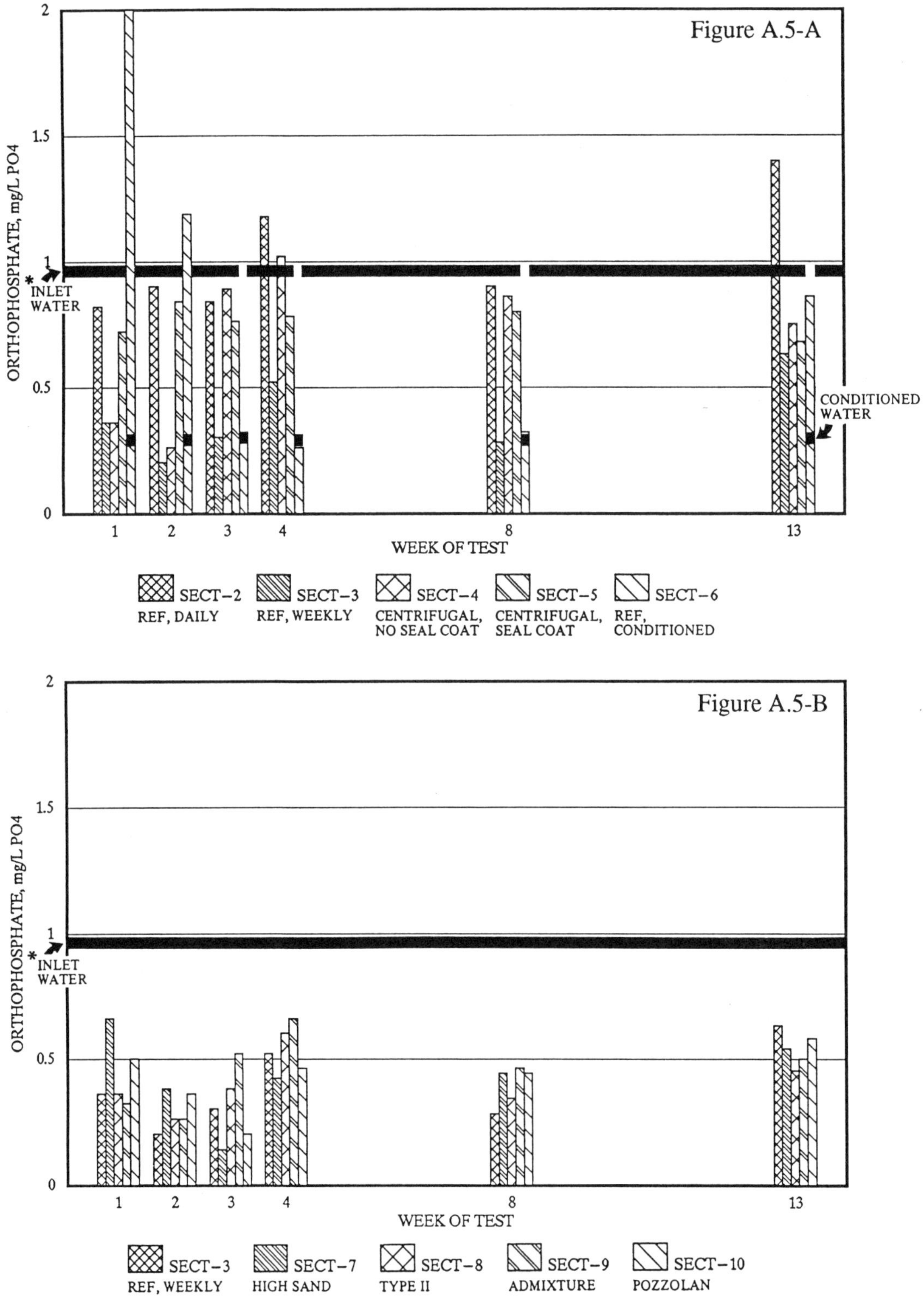

*SECT–1 (Reference Mortar, Continuous Flow) had the same composition as the Inlet Water.

Figure A.5 Marin MWD orthophosphate results

In Situ-Lined Pipe

Allentown Cement Company, Inc.
Evansville, PA

Mill Test on Type I Cement

Evansville Plant — Typical Mill Test

Consigned to: ____________________

B/L No. ____________

Date Shipped ____________

Fineness (Blaine)	383	M2/Kg	
Initial Set (minutes)			
Final Set (hours)			
Vicat (minutes)	110		
Autoclave Expansion	0.06		
Compressive Strenth	PSI	/	MPa
1 Day		/	18.6
3 Days	3700	/	25.5
7 Days	4500	/	31.0
28 Days		/	36.5
Air Content %	8.0		
C 265 (grams SO_3/litre)			
—325 %	89.0		

Silicon Dioxide (SiO_2)—%	20.5	C3S—%	48.1
Aluminum Oxide (Al_2O_3)—%	5.6	C3A—%	11.5
Ferric Oxide (Fe_2O_3)—%	3.0		
Magnesium Oxide (MgO)—%	2.8	Tot Alk—%	0.87
Sulphur Trioxide (SO_3)—%	3.5		
Ignition Loss—%	1.2		
Insoluble Residue—%	0.25		
Calcium Oxide (CaO)—%	62.5		

Test data shown above is representative of silo from which cement was taken. We hereby certify that the cement shipped in above (car–truck) conforms to present standard specifications of ASTM C150-89.

Louis A. Jany
Quality Control Manager

Figure A.6 Mill test results on in situ-lined pipe —Type I cement

In Situ-Lined Pipe

Allentown Cement Company, Inc.
Evansville, PA

Mill Test on Type II Cement

Evansville Plant — Typical Mill Test

Consigned to: ____________________

____

B/L No. ____________

Date Shipped ____________

Fineness (Blaine)	389	M2/Kg		
Initial Set (minutes)				
Final Set (hours)				
Vicat (minutes)	155			
Autoclave Expansion	0.05			
Compressive Strenth	PSI /	MPa		
1 Day	/	16.5		
3 Days	3600 /	24.8		
7 Days	4400 /	30.3		
28 Days	/	36.5		
Air Content %	7.0			
C 265 (grams SO_3/litre)				
—325 %	89.0			
Silicon Dioxide (SiO_2)—%	20.7		C3S—%	54.6
Aluminum Oxide (Al_2O_3)—%	4.2		C3A—%	4.4
Ferric Oxide (Fe_2O_3)—%	4.0			
Magnesium Oxide (MgO)—%	3.0		Tot Alk—%	0.85
Sulphur Trioxide (SO_3)—%	3.0			
Ignition Loss—%	0.9			
Insoluble Residue—%	0.20			
Calcium Oxide (CaO)—%	62.5			

Test data shown above is representative of silo from which cement was taken. We hereby certify that the cement shipped in above (car–truck) conforms to present standard specifications of ASTM C150-89.

Louis A. Jany
Louis A. Jany
Quality Control Manager

Figure A.6 Mill test results on in situ-lined pipe —Type II cement

NATIONAL CEMENT COMPANY, INC.

Southeast Division

FACTORY-LINED PIPE

MILL TEST REPORT

CONSIGNEE:____________________ DESTINATION:____________________

SPECIFICATION: TYPE II ASTM C-150 M-85

DATE SHIPPED	B/L NUMBER
5/27/91	7275

RESULTS OF TESTS

		TIME OF SET (Gilmore)	
Silica:	21.0		
Aluminia:	4.0	Initial Set Time: Minutes	140
Ferric Oxide:	3.1	Final Set Time: Minutes	265
Magnesia:	2.6	325 SIEVE % Retained:	4.4
Sulfur Trioxide:	2.5	Wagner Sq/Cm per gram:	2260
Ignition Loss:	1.2	Blaine Sq/Cm per gram:	3889
Insoluble Residue:	0.42	Autoclave Expansion:	0.03
Alkali as Na2O:	0.52	Air Content:	9.0
POTENTIAL COMPOUNDS		COMPRESSIVE STRENGTHS	
Tricalcium Sicilate:	62.9	3 Day:	3865
		7 Day:	4735
Tricalcium Aluminate:	5.6		

I hereby certify that the cement referenced herein has been tested in our laboratory with the above results.

M. J. Royce
Quality Control Manager

2000 SOUTHBRIDGE PARKWAY, SUITE 620 • BIRMINGHAM, ALABAMA 35209 • AREA CODE 205 • 870-7680 • FAX (205) 870-5777

Figure A.7 Mill test results on factory-lined pipe